AUTOMOTIVE ELECTROMAGNETIC COMPATIBILITY (EMC)

AUTOMOTIVE ELECTROMAGNETIC COMPATIBILITY (EMC)

Terence Rybak

Mark Steffka

KLUWER ACADEMIC PUBLISHERS
Boston / Dordrecht / New York / London

Distributors for North, Central and South America:
Kluwer Academic Publishers
101 Philip Drive
Assinippi Park
Norwell, Massachusetts 02061 USA
Telephone (781) 871-6600
Fax (781) 871-6528
E-Mail: <kluwer@wkap.com>

Distributors for all other countries:
Kluwer Academic Publishers Group
Post Office Box 322
3300 AH Dordrecht, THE NETHERLANDS
Telephone 31 78 6576 000
Fax 31 78 6576 254
E-Mail: <services@wkap.nl>

Electronic Services <http://www.wkap.nl>

Library of Congress Cataloging-in-Publication Data

Automotive Electromagnetic Compatibility (EMC)
Terence Rybak and Mark Steffka
ISBN 1-4020-7713-0

Printed on acid-free paper.

Printed in United Kingdom by Biddles/IBT Global

Contents

Preface

Anyone who has operated, serviced, or designed an automobile or truck in the last few years has most certainly noticed that the age of electronics in our vehicles is here! Electronic components and systems are used for everything from the traditional entertainment system to the latest in "drive by wire", to two-way communication and navigation. The interesting fact is that the automotive industry has been based upon mechanical and materials engineering for much of its history without many of the techniques of electrical and electronic engineering. The emissions controls requirements of the 1970's are generally recognized as the time when electronics started to make their way into the previous mechanically based systems and functions.

While this revolution was going on, the electronics industry developed issues and concepts that were addressed to allow interoperation of the systems in the presence of each other and with the external environment. This included the study of electromagnetic compatibility, as systems and components started to have influence upon each other just due to their operation. EMC developed over the years, and has become a specialized area of engineering applicable to any area of systems that included electronics. Many well-understood aspects of EMC have been developed, just as many aspects of automotive systems have been developed. We are now at a point where the issues of EMC are becoming more and more integrated into the automotive industry.

Unfortunately, the auto industry and the EMC discipline have not much interacted with each other, except for special cases involving specific groups or people that worked in the field. This has meant that there are vast numbers of automotive engineers and technicians without an understanding of EMC, and many specialized and competent EMC professionals that are not experienced in the automotive industry.

The solution to this problem? A body of knowledge that can put, in automotive terms, the concepts and issues in EMC. This is what the authors of this text have attempted to do. This book is intended to be a "one-stop" reference and introduction to the subject of automotive EMC that will enable those working in the auto industry to be able to identify EMC issues, causes, and corrective actions, as well as provide references for those wishing to research the subject further.

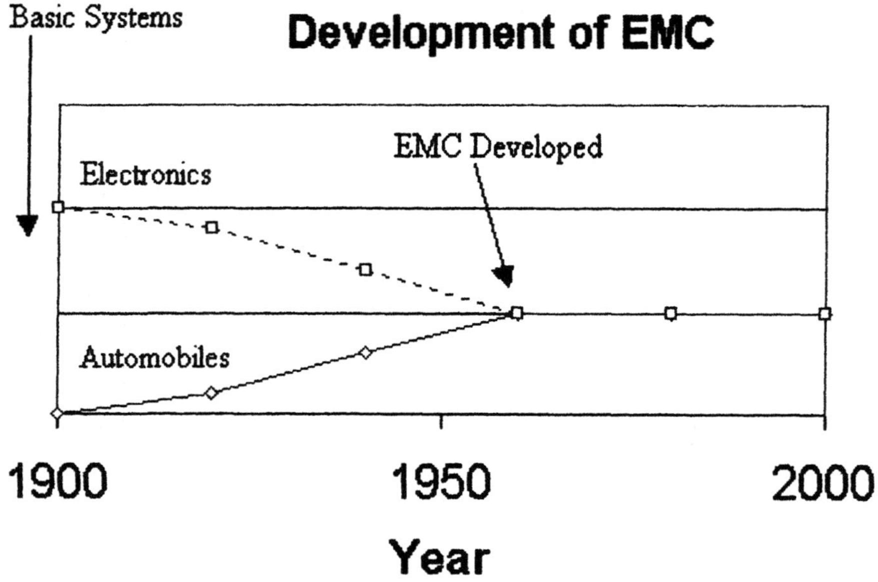

Figure 1. **Convergence Of Growth In Electronics And Automobiles**

The format of the text is intended to facilitate either an introduction to the subject of automotive EMC or as a basis for deeper research. This is accomplished by breaking the material into chapters that are related to specific automotive issues, and then providing the EMC background to those issues. The description of those chapters is:

Chapter 1 discusses the evolution of EMC and how it's emerged in relationship to technology. It also describes some of the first issues in the area of automotive EMC and the impact that solid-state devices have had upon automotive EMC. There is also a description of the current issues, and a forecast of future issues.

Chapter 2 establishes the basis of components versus systems. There are discussions on the importance of component level and vehicle level systems testing, and comparisons to tests that are conducted in other industries.

The subject of "power and signal integrity" is presented in Chapter 3 as a concept to contrast with the common approach of "power and ground". This has been used for a long time and can be a confusing approach when trying to work on EMC problems and issues

Within many undergraduate programs resides limited understanding of basic antennas, transmissions lines, and passive components (inductors, and capacitors), which causes much concern and confusion. Undergraduate engineering programs have moved to computer engineering, with the result that studying radio frequency components and basic items have been deleted on the belief that they are unnecessary. These are important items to know for any EMC study, and the material is covered in Chapter 4.

A study of EMC must include covering the fundamentals of electromagnetic field theory and the physical laws. Maxwell's equations are reviewed as to their applicability to EMC issues, and to provide a framework for understanding the physics of the issues. Also included are the concepts of near and far field's, measurement of field strength, and propagation characteristics. Path loss is included to provide insight into the attenuation of signals that need to be considered in immunity issues. This material is presented in Chapter 5.

Chapter 6 is an overview of test methods as related to vehicle-level testing and component-level testing. While some component-level testing methods are standard across industries, and the unique aspects of using test equipment for vehicle level testing is discussed.

EMC modeling is a method that is evolving as a method for the future. An overview of current tools and possible use of those tools is covered in Chapter 7.

Chapter 8 discusses effects of cabling and harnesses used to connect electronic modules and sensors systems (most of today's vehicles). These may include electronic systems for engine operation, vehicle control, or entertainment systems and so forth. The automotive industry has addressed compatibility issues through EMC departments designed to resolve and address problems and develop solutions in the component design phase by working with suppliers of those particular components in integrating them into the vehicle systems. In addition, the manufacturers conduct extensive testing to verify both the component- and vehicle-level EMC performance before a vehicle is offered for sale.

Automotive electrical and electronic systems have unique characteristics. Some components operate on low voltages and current levels, while high voltage and high current systems are used throughout the vehicle. These include the ignition system, alternator and charging system, and other high current and high-voltage devices. As we move forward into more complicated vehicle system architectures and electronics, we will see that EMC will become more important as an item to consider in vehicle engineering. There is discussion in Chapter 9 of the current and future data communication systems and networks on vehicles, with examples of EMC issues associated with those systems.

Several standards, rules, and regulations cover the automotive industry both from industrial and regulatory standpoints. These include the directives and requirements of international bodies such as the European Union and the Canadian government. In the United States, the FCC has responsibility for the control of radiated emissions and interference for products, although the automotive industry has some exemptions from these requirements. Chapter 10 discusses the requirements that do exist and practices that are incorporated where requirements or regulations do not exist.

A challenge as we move forward in automotive EMC is that of controlling and understanding the vehicle level electrical transients that occur. Many of these effects are just now becoming more frequent, and the difficulty is understanding the sources of these transients. Chapter 11 will review the research and current development of quantitative methods of defining vehicle level transients.

Electrostatic discharge is an area that also merits consideration in the automotive system. This is because there are many devices that can be sensitive to ESD. Chapter 12 overviews ESD, its nature, and the test methods that are used in the automotive industry.

Is this text intended to be the first and last source of all material on automotive system EMC? Absolutely not! It has been the author's intention to provide only a starting point in this subject. We're sure that as time goes by, others will meet the challenge of this discipline and create information that keeps up with the auto EMC industry. It is with that anticipation that this work is written.

Chapter 1

What is EMC?

1.1 BACKGROUND

This book is a study of the issues, experiences, and trends in automotive system electromagnetic compatibility (EMC). EMC and automotive systems is an area evolving from the early days of few electrical devices to the highly complex electronic components in vehicles. This book will look at how the EMC of automotive systems has become a major issue, and describe the tools and techniques of automotive systems EMC. We will look at various system components architectures and EMC issues that are associated with those systems.

We'll assume the reader has a basic understanding of automotive electrical and electronic systems. Let's begin this book by defining EMC as "the ability of an electronic system to function properly in its intended electromagnetic environment, and to not contribute interference to other systems in the environment." The goal is to have the electronic system be immune from the emissions of other systems, not interfere with the operation of other systems, and not interfere with its own operation.

The basic model of EMC can be thought of as shown in Figure 1.1. Let's assume we have device A and device B. Our goal in EMC is to have A and B operate in the presence of each other as well as operate in the presence of external environments. We do not want A to interfere with B nor do we want to B to interfere with the operation of A. We also do not want to the external environment (for example, radio broadcast transmitters) to cause either A or B to operate incorrectly.

A key concept in EMC is the "source-path-receiver" relationship, which is shown in Figure 1.2.

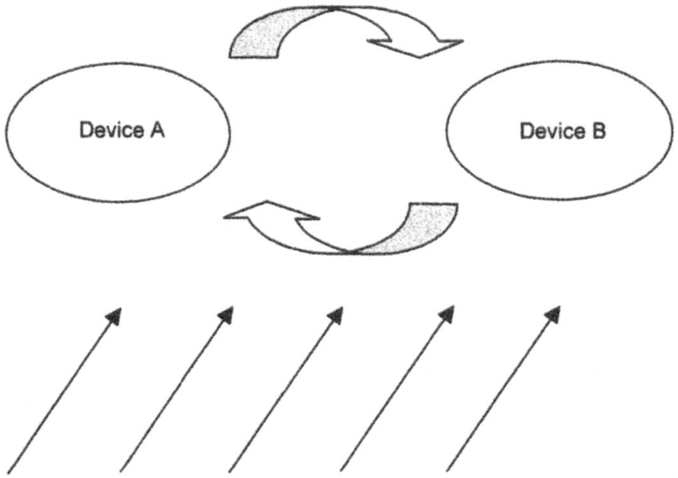

Figure 1.1. **Basic EMC Model**

Figure 1.2. **Source-Path-Receiver Model**

This shows that fundamentally, we have three basic elements comprising EMC; the first is the source, the next is the path, and the final one is a receiver. Receivers may be of two different types; "intentional" or "unintentional." An example of an intentional receiver would be a radio or television receiver, and an example of an unintentional receiver would be a computer or some type of electronic device. This is the basic model that we use in addressing EMC problems, and one to which we always attempt to reduce our EMC problems.

How can we ensure EMC with knowledge of this model?

- We could suppress the energy at the source (meaning that we could reduce the amount of energy being radiated).
- We could address the path itself; this path might be conducted through a wire, or radiated through the air.

- We could address the receiver's characteristics and make it a "hardened" receiver.

A key issue with regard to receivers of energy is the concept of immunity and susceptibility. In the automotive industry, we use the term immunity, whereas most other areas that are concerned with the EMC disciplines use the term "susceptibility". For purpose of this book we will define susceptibility and immunity as the related as shown in Figure 1.3. If we move up the vertical axis, we have increasing levels of susceptibility; and if we look to a horizontal axis and move to the right that is decreasing immunity. What we're saying is that if we have a susceptible component or system, we have little immunity; and something that is immune has little susceptibility. This is not unlike the human body immune system; if we say that we have a highly effective immune system, we are less susceptible to illness.

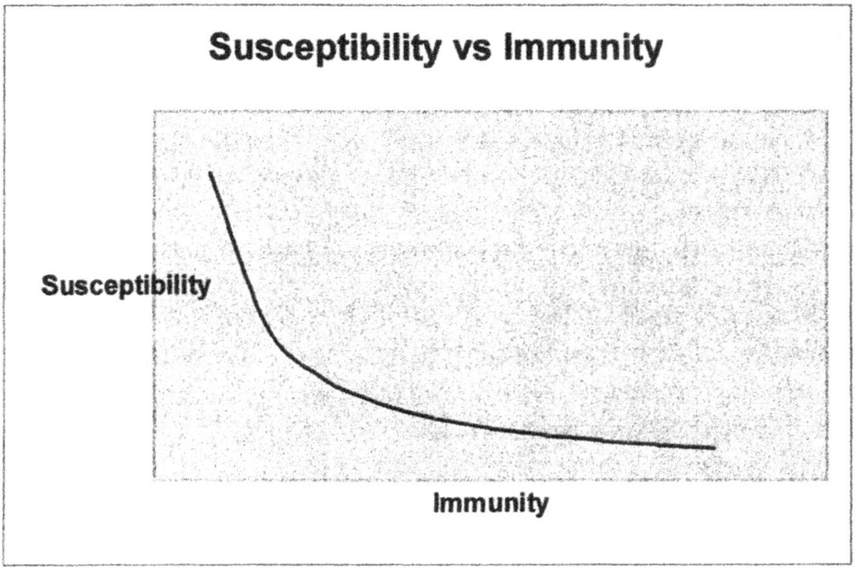

Figure 1.3. **Susceptibility and Immunity**

1.2 TECHNOLOGY AND EMC

It is important to understand the origin and the migration of EMC issues which have developed as a result of technological innovations during the last century. At the beginning of the 20th century, the technology of the day with regard to communications consisted primarily of generating high frequency radio signal by generation of high-voltage discharges.

The systems were used to send short messages and allowed experimentation with propagation characteristics and high-frequency transmissions. Since there were few of these systems, there were few instances of EMC.

In the late 1800's to the early 1900's, there was much work done with "wireless" (the original term for radio) communication. One person heavily involved in this work was Gugliemo Marconi. Marconi did many experiments in Italy and was interested in how to send messages across the airwaves. He studied the experiments that Hertz and other pioneers in the field were performing, and he wanted to conduct such experiments himself. His work involved sending RF energy across distances using some of the techniques shown in Figures 1.4 and 1.5. The key elements in his experiments were a method to create RF energy (which was accomplished by a high voltage coil), a power source (battery), and a way to transfer the energy to the air (plates). Today antennas accomplish this energy transfer. He then constructed a way to receive the energy by creating a receiver system that consisted of plates and a device that allowed the detection of the spark energy allowing the sound to be heard in headphones. A schematic diagram of his original system is shown in Figure 1.6. A schematic of one of his later developments is shown in Figure 1.7.

Figure 1.4. **Marconi's 1895 Transmitter**

Reproduced with the permission of Marconi Corporation plc from www.MarconiCalling.com

Figure 1.5. **Marconi at Age 22 With His First Patented Wireless Receiver (1896)**

Reproduced with the permission of Marconi Corporation plc from www.MarconiCalling.com

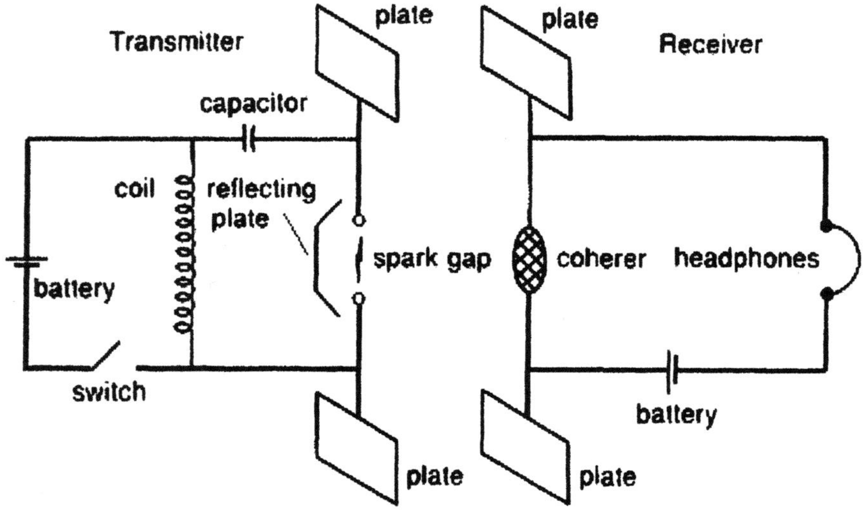

Figure 1.6. **Schematic of Marconi's Original Wireless System**

Reproduced with the permission of Marconi Corporation plc from www.MarconiCalling.com

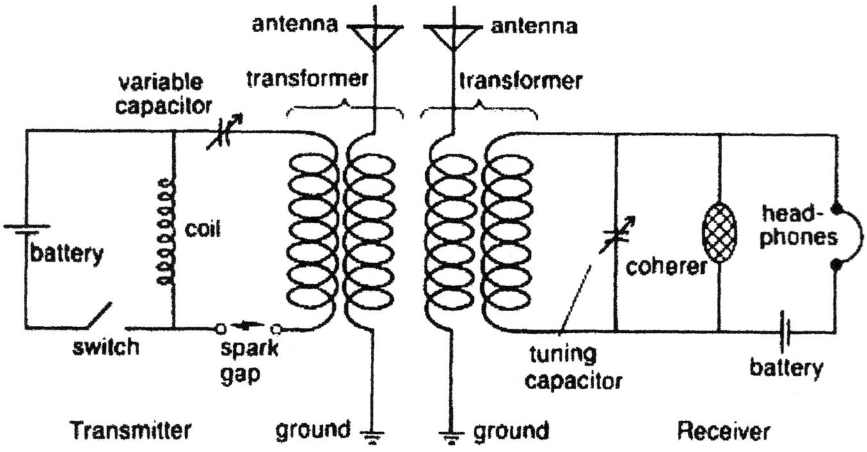

Figure 1.7. **Detailed Schematic of Marconi's 1900 Wireless System**

Reproduced with the permission of Marconi Corporation plc from www.MarconiCalling.com

1.3 COMMUNICATION TECHNOLOGY EVOLUTION

If we look at the early days of electronic communication, we see that there was at first a revolution, caused by spark-gap technology, and there has subsequently been evolution. The EMC issues associated with spark technology were few because there were few receivers. Although the nature of spark-gap transmitters would be that the signals would have a wide bandwidth in general, people could identify which transmission they wanted to receive. There were minimal EMC issues.

As technology evolved and there existed a need for more complex receivers, additional components were developed. The major advance in technology was the development of vacuum tubes. This was a major revolution, as it finally allowed amplification of low power circuits, impossible with passive components. Fortunately, the characteristics of vacuum tubes are that they require large amounts of power and high-voltages, which minimized potential EMC issues.

The next major step in the technology evolution was "solid-state" (so called because these were solid material and did not require fragile class envelopes and vacuums), primarily developed after WWII. The first devices consisted of diodes and transistors, which were intended to replace the already many applications of vacuum tubes. One characteristic of solid-state devices has actually increased EMC issues; solid-state devices require low energy and low voltage levels to operate. By their nature solid-state devices operate on low-level signals, and can be affected by low-level amounts of radio frequency energy. Of course, today's systems include many solid-state devices, and we've moved now to areas of high-scale integration and small geometry with even lower energy levels required to operate those devices. An additional item that has been developing recently is a high demand for radio-frequency spectrum (evidenced by the number of auctions that the FCC [Federal Communications Commission] conducts) as the demand for wireless communications systems increases. This additional aspect has placed an awareness of EMC because of many communications devices that could be affected by EMC issues.

In summary, the combination of the technology evolution and the demand for RF spectrum has created many EMC issues that exist today. This is shown in Table 1.1: Observe the left-hand side which shows the migration from spark to tubes to solid-state technology, and then as the number of signals and emissions increase, the number of EMC issues increase, with possibly no end in sight.

Table 1.1 **Increasing Numbers of Signals Results in a Greater Number of EMC Issues**

	Technology	EMC Issues
Revolution	Spark	Few - Not many receivers
Evolution	Vacuum tube equipment	Some - Tubes require significant power
	Solid state	Many - Low level signals, Low level operation can be affected by RF
▼	Spectrum utilization is high	

1.4 CONVERGENCE OF TECHNOLOGY AND AUTOMOTIVE SYSTEMS

Why is EMC important in any automotive environment? In recent years, automotive systems have increased their content of both electrical and electronic devices. Electronic systems on today's vehicles include many functions that were not incorporated on previous vehicles, or even imagined! For example, today's systems contain active electronics; and many electronics modules include microprocessors, transistors, and switching devices. These electronics provide increasingly more control functions on vehicles. The concern is that these assemblies and components may emit energy and will also react to external sources of energy, resulting in unanticipated performance of these vehicle systems.

Why study EMC? There are a number reasons. The first one is to understand how to meet various legal requirements for EMC in different countries. For example, the European Union and Canada both have requirements for radiated emissions. In the United States, emissions from electronic devices are regulated by the FCC; however, what is unique about the automotive industry is that the devices on automotive systems are exempt from the radiated emission requirements included in "Part 15" of the rules and regulations, provided they do not cause "harmful interference."

Another reason might be meeting product requirements. Clearly consumers now are very demanding regarding system and component performance. Safety, cost of retrofitting "fixes", and legal issues are also incentives to ensure EMC.

Today, although many of the components that were used on early automobiles (high voltage distributors) may still be used, we now have extensive use of digital devices on all kinds of systems including automotive systems. You may recall from Fourier series analysis that one of the aspects of digital devices (which operate at square waves) is that a square wave contains many harmonics. These harmonics are based upon the (10-90 percent) rise time and the fall time of the square wave signal as shown in Figures 1.8 and 1.9. We will discuss later in detail the EMC challenges of these.

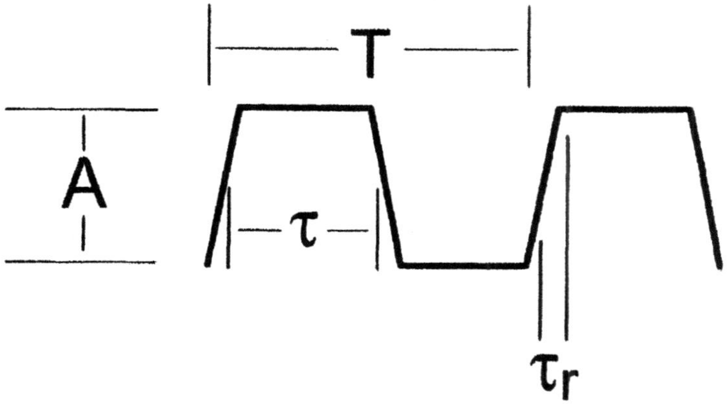

Figure 1.8. **Square Wave**

As shown in Figure 1.9, we have a frequency content in the rise and fall edges that is much greater than the fundamental frequency, and a frequency content of zero in the areas where the signal is at a level value. Digital devices are used on many systems today and we will use more in future applications. This is one of the major sources of EMC (specifically emissions) that we experience in today's automotive industry.

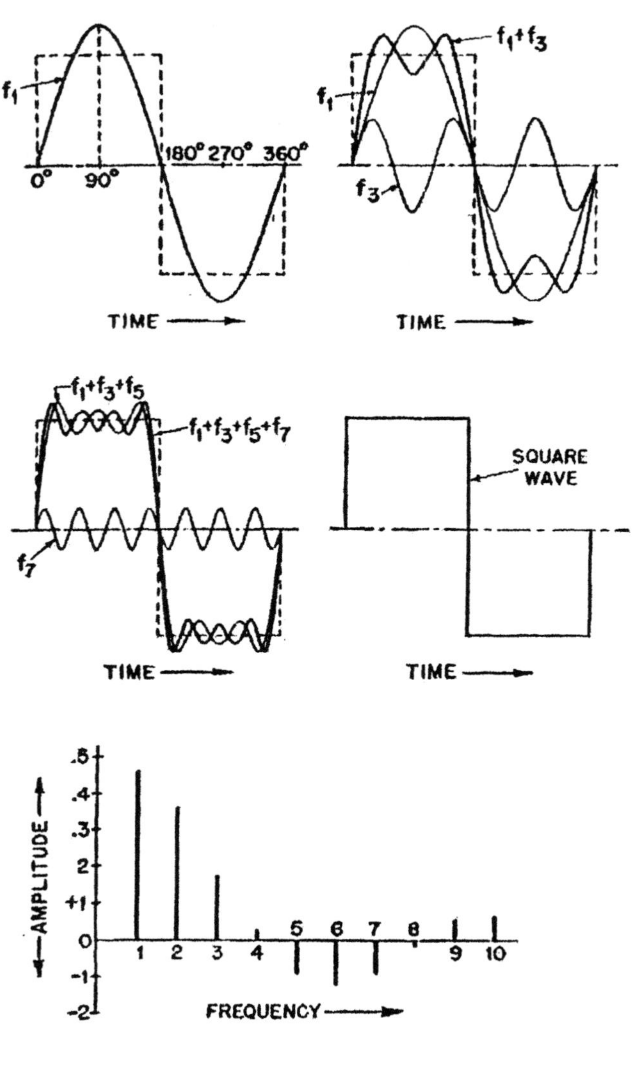

Figure 1.9 **Contribution of a Square Wave to Various Frequency Spectra**

1.5 FUTURE TRENDS

The importance of EMC to product design and manufacturing will probably increse due to a number of reasons.

- In the area of computer technologies, we have seen that the clock speeds are moving higher and higher. The challenge with this technology is that as the speeds increase, cases, structures, enclosures, etc... start to have minimal effect with respect to their shielding capabilities. They can even act as fairly efficient radiators.

- There is also much work being done in the area of extremely short wavelength communication for high-speed wireless digital communications. The EMC aspects of this technology will need to be characterized.

- One is due to the utilization of new technologies, such as "Ultra Wide Band" (UWB). This is a low power and short-range radio system that will operate at approximately 2 – 3 GHz. The system is designed to be robust against interference, the levels of robustness that will be needed once there is significant implementation of the systems, will be evident.

- As we will see in a later chapter, from an antenna perspective, an effective radiator is a one-quarter or a one-half- wavelength. At 600 MHz, a one-half-wavelength is only 24 cm (about 10 in.). This may require a better design for emission suppression and immunity. In other words, there is no shielding attenuation of a slot or gap when it is half wavelength long; and once an emission from the product gets through a slot, it acts like any other wave in both its near- and far-field effects on surrounding devices.

- The slot is a receiving aperture or antenna and lets in RFI environmental noise. These conditions may cause a focus on EMC be at the printed-circuit-board level, in addition to the EMC issues that may been seen at lower frequencies due to wire harnesses or assemblies.

Combined with the design concerns at these high frequencies, it is also likely that EMC measurement difficulties will increase due to a number of reasons that include:

- The noise floor issues in spectrum analyzers and receivers may have an impact upon attempts to measure low level emissions. Antenna-to-receiver cables become lossier, affecting the ability to measure signals above the noise floor of the spectrum analyzer or receiver.

- Immunity to radiated fields is more difficult because of the narrow beam width and the product's apertures, such as cable ports and slots.
- At high frequencies, the product emissions beams are narrower and are “directional”, which means they are difficult to find.
- The antennas that would be used to measure emissions would also have narrower beam widths, making them difficult to aim at the product and measure the radiated signal.

Given the continual evolution in radio frequency based systems and for the examples just given, it should be seen that EMC will be a concern for many years to come!

Chapter 2

System Level Issues

2.1 DEFINITION OF COMPONENT AND SYSTEM

The study of EMC is required only when systems are involved. There are no EMC issues inherently associated with a switch, a spark plug, or even a microprocessor. These are components. EMC issues occur when components are operating and interface with the external environment. This will be explained in the following text.

What is a component? What is a system? We need to understand what each of those is. For the purposes of this text, we are going to define a component as an entity that can be described as having some physical size and mass. A transistor is a component; an engine controller is a component. We can describe the size and mass of the transistor, and perhaps some of its other characteristics, including number of leads, color or shape. The same can be said about an engine controller, or an automobile. We can "hold a component in our hands", such as the transistor, controller, or if our hands were big enough, and we could hold the planet Earth in our hands. We will also establish that an assembly is comprised of components. Transistors are made of semiconductor material, plastic and metal, and wires. Controllers are made of a printed wiring board, active and passive components, and connectors. In our definition, we will state that components and assemblies are elements of systems. See Figure 2.1.

What are systems? We will define a system as the interaction of components with one another and the external environment. We can say the transistor is an element of the "amplification system". This means that we have an environmental interaction (the incoming signal, transistor operation, and then the energy sent out (the amplified signal).

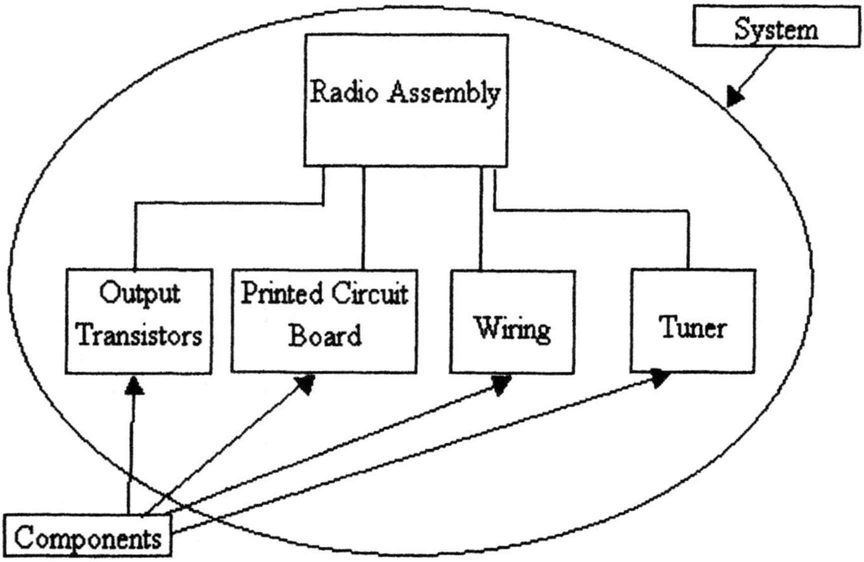

Figure 2.1 **System and Components**

Thus we can see that according to our criteria, we can identify a component, and the system only exists when the components are operating. Let's look at another example of an automotive system. The ignition "system" includes a number of elements (components, energy delivery, and interaction with the environment). The system only exists when a spark is created that causes combustion. When the vehicle is at rest and the engine not running, all we have is collection of components.

Sometimes the term "sub-system" is used. This is a term that is not readily understood. Perhaps it is used to mean a portion of the components that are in a system? If that's the case, then it appropriate to use the term assembly to identify the physical nature of the components. Sometimes sub-system is used to identify particular systems of an overall system. An example of this is the "entertainment sub-system" of the automotive system. In this text, we will either use the term component, or system. Use of the word system will imply the operation of a number of components, and an interaction with the environment

2.2 SIGNIFICANCE TO EMC

The reason this is an important point is that we will see that many EMC issues result from system-level characteristics, rather than inherent deficiencies with components. This condition is especially true in the automotive industry, perhaps more than in other industries, due to the physical sizes of the components and the functions of the systems. We will review the attributes of components and how those attributes contribute to decreasing the ability to achieve a desired EMC from a system, or how they contribute to increasing the ability to reach a desired EMC state.

One of the key issues in EMC is recognizing that the process of attaching wires and conductors to electrical components creates a system that may have characteristics different from just the electrical component itself. This is due to the energy that can be transferred as a result of the wiring. People working on EMC issues often lose track of this fact, and try to "force" component-level performance into a description of system-level constraints. This text will emphasize that this can cause much confusion!

Chapter 3

Power and Signal Return

3.1 INTRODUCTION

Many automotive EMC problems are attributed to "bad ground" connections. Bad ground seems to be the cause of many problems in all types of electrical circuits. The reason that there are bad ground connections is simple. There is not a "ground" anywhere on a vehicle! The reason there is no ground connection is also simple. The vehicle is intended to travel <u>on</u> the ground, not <u>attached</u> to it. Actually, the one time when there can be a ground connection on a vehicle this is shown in Figure 3.1:

Figure 3.1. **Vehicle With Ground Connection**

In this case, if the ground connection is maintained, it can be seen that the vehicle is of little use as a transportation method if it can only travel as far as the ground cable allows it.

The use of the term ground unfortunately has become used to describe the path where the return currents are assumed to be flowing. As a matter of fact, there is a circular definition of the term electrical ground. Many writings on electrical circuits refer to the ground as the "sink of the power or signal currents". That definition MAY be satisfactory if the return currents read *this* definition and then *consult with the circuit designer* to find out where they should be flowing! There are interesting definitions that have been developed. One heard recently was the concept of "dirty" and "clean" grounds. This emphasizes the fact that a "ground" is not what it is supposed to be, since we seem to keep making more definitions when realize when our existing ones do not work!

It is correct to look at the path of the return currents as the "return". Doing so will eliminate the underlying assumptions about ground connections that are not always true. For example, it is sometimes assumed that the ground is a zero impedance path and can sink infinite amounts of current. The problem is that in the real world, there is no such thing as zero impedance and there is also a limitation on the current carrying capability of the return path. The other problem with referring to "ground" connections is that there are at least three uses of the term "ground". We will discuss these later. It is the authors' intention to never use the term "ground" in this text when power or signal return path is actually meant. (However, some old habits are difficult to eliminate!)

Let's look at some basic facts in order to develop our concept of return rather than ground.

- The first concept is that every current must return to its source. This is a fact of nature. If this were not true, there would be pools of charge created by the accumulation of current flow, which does not happen.
- The next item is that the majority of the current takes the path of least...IMPEDANCE. We're sure many of you learned current takes the path of least "resistance". That is a true statement – when we are dealing with low frequencies or D.C. Once we have a frequency greater than D.C., which is nearly all the time (including "pulsed" DC, which has a period near zero), then we need to understand that impedance is important.
- The last key point is that in order to understand the circuit, the current source and return must be known for each current. If they are assumed to be on the same line, and it is not understood where the currents flow, this can lead to difficulty in creating a model of the actual conditions.

In summary:

1. There is no ground connection on a vehicle (or for any electrical circuit that does not have a wire or cable going from the circuit to the earth).
2. Current takes the path of least impedance.

3. Frequencies greater than DC means that currents will flow different from what was assumed to be the path in the DC current flow.

3.2 CURRENT PATH

Let us now look at the impact of these statements. What they imply is that, if the current has a frequency greater than D.C., then the concept of impedance must be considered. This means that the current paths may be defined by either the inductance or capacitance of the circuit, NOT ONLY THE RESISTANCE!

If the current path is defined by the inductance of the circuit, then a major contributor is the size of the current loop. This will be discussed in more detail later. Note: implicit in this definition is the actual current loop – not the assumptions about the wiring harness, since the harness may not always be conducting the current it is assumed to be conducting. If the impedance is defined by the capacitance of the circuit, this is due to the relative location and spacing of the conductors, which are clearly not the DC circuit paths.

Least resistance may not be equal to least impedance! Many times in solving EMC problems at the circuit board level there is the incorporation of a "ground plane". This again is an example of using confusing terminology. The purpose of the plane, which normally consists of a conductive surface over the majority of the area, is to allow the current to define its own "least impedance" path back to the source. What is significant is that this path may even be the path of higher resistance, yet lower impedance! This would seem to contradict "common sense"! See Figure 3.2.

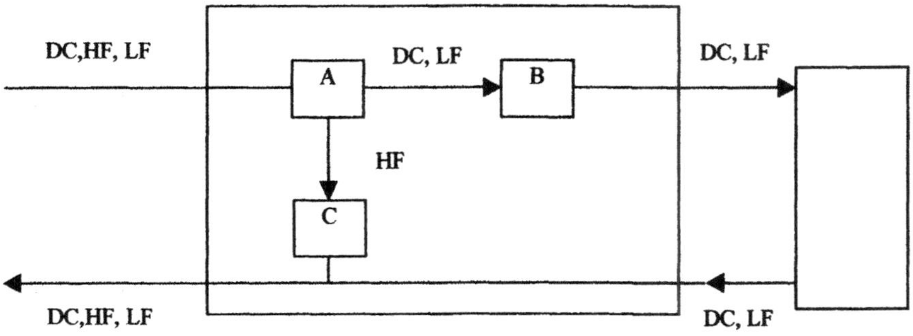

Figure 3.2. **Circuit Board Showing Different Paths of Return Current Flow**

There are some conditions where it is appropriate to refer to the "ground" connections. These are generally related to safety considerations and primary power in residential and/or commercial installations. In this case, there are connections that routed back to a rod that is driven into the ground (earth). The purpose of this is to provide an alternate path for the current to flow in the event of a circuit fault. This ground connection is the third pin on the three-prong electrical connectors that are in use today. Along with the ground connection, today's electrical codes require that there be a "polarity" to the connection. This is also intended to protect the operator from a safety issue or concern. Photos of the three-prong connector and the polarized connector are shown below, with the connections labeled.

In the electrical codes, there are also reference to the "hot" and "neutral" connections. Figure 3.3 also shows which lines connect to which terminals..

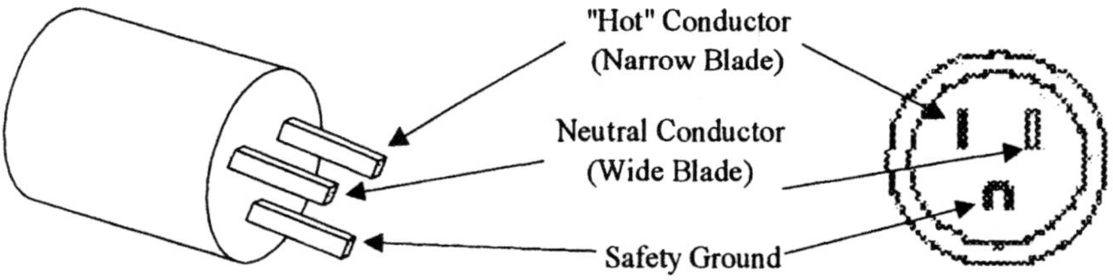

Figure 3.3. **Three Pin Electrical Plug Incorporating Safety Ground**

There may be a second meaning of the term grounding – this is typically used in electronic circuits. This may actually be a voltage reference, where the current in the voltage reference line is very near zero. This is shown in Figure 3.4.

This type of connection should not be called grounding – it should be called voltage reference, because that is the function it is performing.

Let's now look at another concept that is frequently used, and see if we can better define the actual conditions that are taking place. These are shown in Figure 3.5., and should be called single and multi-point return connections.

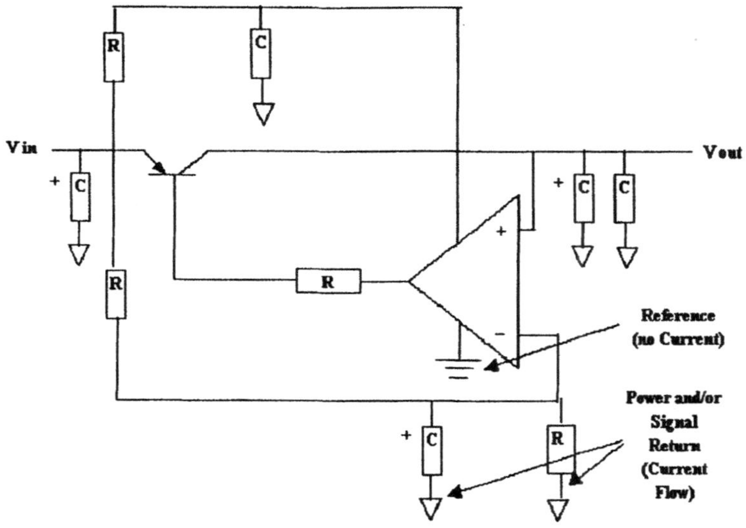

Figure 3.4 **Voltage Reference "Ground"**

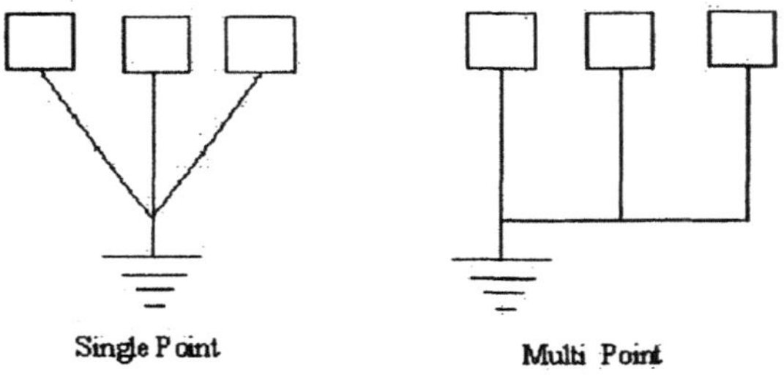

Figure 3.5. **Multi Point And Single Point Return**

What is interesting about these two diagrams is that they try to bridge between both the real world and the ideal world. What we mean by this is that the connection scheme would seem to indicate that the wiring is different between the two configurations. What is significant is that, in the multi-point configuration, if the impedances of the line between the elements are very low, then the connections would or could be represented by the signal point connection. Therefore, it is more correct to insert some impedance in the lines that connect the elements. Once this is done, it then becomes apparent what the characteristics of each of the connection methods is.

Table 3.1 **Comparison Of Different Types Of Returns**

	Single Point	**Multipoint**
Advantages	Reference to same point Best for low frequency	Less wire needed
Disadvantages	Coupling between conductors	Must have near zero impedance

In summary, let's look at what we've learned in this chapter.

- The signal ground is not always the signal return path.
- EMC problems are frequently related to assuming that there is not a good "ground".
- It is important to know the paths of the return currents, and that those paths depend upon the impedance of the circuit.

For consistency throughout this book we will use the following notations with their associated meaning, shown in Figure 3.6..

- Safety ground = zero current during normal operation
- Signal reference = near zero current during normal operation
- Signal or power return = current carrying connections

Another problem with the use of the term "ground" is that it has the connotation of multiple electrical points that are at the same potential (0 volts) all the time. Unfortunately this is not true in many situations and can lead to difficulty in diagnosis of various types of problems

Let's look again at understanding the concept of current taking a path of least impedance. By reviewing Figure 3.2, if we have both DC and low frequency in this particular in a circuit, they may both take the same path as shown below. However, if we have some type of high-frequency signal (and high-frequency in this case may actually be on the order of tens of kHz) a

high-frequency current may take another path, which is actually the lesser impedance. This is an example of why DC and AC signals may take two different paths, because the current takes path of least impedance. This again could cause confusion trying to diagnose EMC problems.

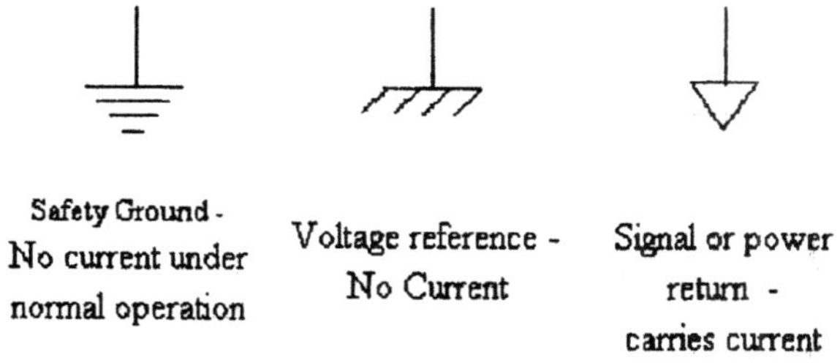

Figure 3.6. **Return Symbols**

3.3 SAFETY GROUNDING

Safety grounding is defined as referencing an electrical circuit or circuits to earth or a common reference plane for preventing shock hazards and/or for enhancing operability of the circuit and EMI control. Bonding is defined as the process by which a low impedance path is established for grounding or shielding purposes. Because the terms "grounding" and "bonding" are often used interchangeably, it leads to confusion. In this section, only the grounding of electrical circuits, not the grounding of metallic components such as electrical equipment cases, cabling conduit, pipes, and hoses (sometimes referred to as bonding), is addressed

Safety grounding an electrical power circuit provides a current return path during an electrical fault. This allows the fuse or circuit breaker to operate properly and prevents shock hazards to personnel. This is accomplished by ensuring that the fault current path has impedance that is small and an ampacity (current carrying capacity) high enough to allow the circuit breaker or other protection device to operate. Additionally, the voltage generated by the fault current between the equipment case and ground must be low enough to meet safety requirements. Voltage generated due to the fault is:

$$V_{fault} = I_{fault} * R_{bond}$$

where I_{fault} is the fault current and R_{bond} is the resistance of the equipment ground connection. This resistance includes the resistance of each electrical bond in the ground connection and the resistance of the grounding strap or jumper used in the ground connection. I_{fault} is the maximum amount of current that the electrical power system can source.

Some electrical circuits require connection to a common reference plane ("ground" plane) in order to operate efficiently. Grounding of filter components and other EMI control measures increases EMI suppression. The line-to-ground or feed-through capacitors used to suppress noise must have a low impedance path to the source of the noise. In order to shunt the currents from line to equipment enclosure (preventing noise from escaping onto power lines), the resistance and the reactance of the bonds in the path between noise source and line-to-ground capacitor must be sufficiently low over the bandwidth at which the line-to-ground capacitors operate. It is important to remember that grounding is not a "cure-all" for EMI and improper grounding may aggravate noise problems. In regard to EMI control, the objectives of a good grounding scheme are to minimize noise voltages from noise currents flowing through common impedance and to avoid ground loops.

Figures 3.7 to 3.9 are schematics of isolation for current loops. The single reference ground is a commonly used grounding concept for aerospace projects. The aim of the single point and single reference ground is to reduce low frequency and dc current flow in the ground plane. Adding to the grounding confusion is the fact that the term "single point" may be used to refer to a single point star or a layered single point ground. For consistency, a single point star ground is referred to as a star ground and layered single point ground is referred to as a single point ground. Additional information on grounding schemes is found in references. It is important to remember that one type of ground scheme can be utilized for power signals, another for RF signals, and yet another for analog signals and cable shields. It is important to utilize the various concepts as needed to meet the requirements of safety, enhanced operability, and EMI control.

3.4 SINGLE POINT GROUND (SINGLE REFERENCE)

The single reference ground scheme is a derivative of the star ground. Each isolated electrical system is referenced once to the ground plane. In most cases, the ground plane is the vehicle or payload carrier structure. The short jumpers used to reference to ground locally and the metallic structure between the grounding points (if good bonding practices are implemented) have a lower impedance than a wire or cable used to reference the isolated systems in a star ground. This lowers noise voltages caused by noise currents flowing in the ground system.

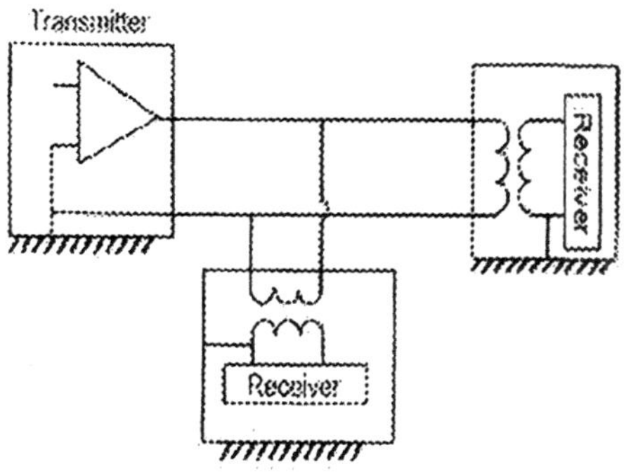

Figure 3.7. **Data Bus Isolation**

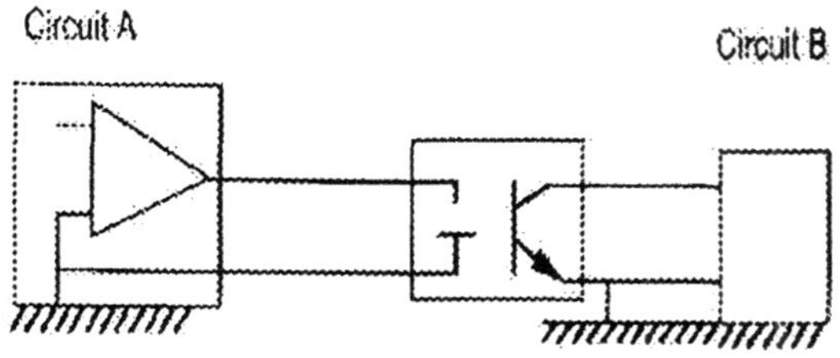

Figure 3.8. **Optical Isolation**

Ground Loop Isolation

It is important to maintain isolation to avoid single point ground violations. These violations result in ground loops that radiate noise or pick up noise from outside sources. In an electrical power distribution system, a switched-mode power supply with transformer isolation is used to prevent ground loops. The power supply output is referenced to ground and any loads powered by the supply are isolated from structure. A power supply in one box provides electrical power to a second box. The input of the second box is isolated from ground. Signals sent between boxes can be isolated in a number of various ways. The most common methods are transformer isolation, optical isolation, balanced differential circuits, and single-ended circuits with dedicated returns. Figure 3.7 shows a control line using optical isolation. Figure 3.8 shows a balanced differential data line between two boxes. Another option is a single-ended circuit in which current is returned on a dedicated wire instead of the ground plane.

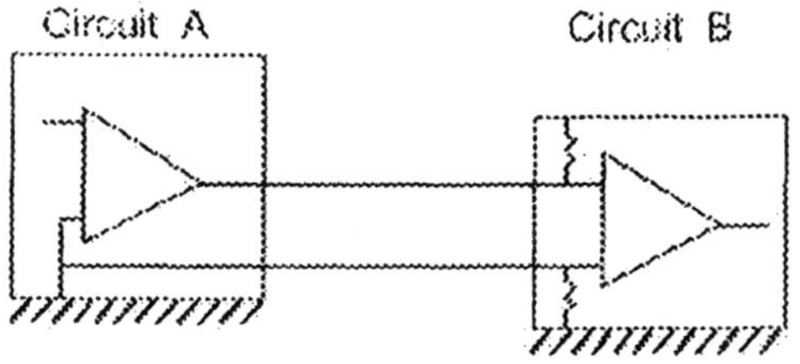

Figure 3.9. **Balanced Differential Data Lines**

The ideal way to prevent common-impedance coupling is to use separate returns for each circuit. Since this is not always possible, careful planning of the circuit layout is needed. Figure 3.10 is a schematic of a good rule of thumb to use when sharing returns. Place quiet circuits farthest from the single point ground and the noisy circuits closest to the ground connection. This limits the common-impedance coupling by limiting the impedance of the return path for the noisy circuit. The inverse of this is to place the circuits that are insensitive to common-impedance coupling farther away from the ground connection than the sensitive circuits. The closer the circuit is to the ground point, the smaller the shared impedance to cause a noise voltage.

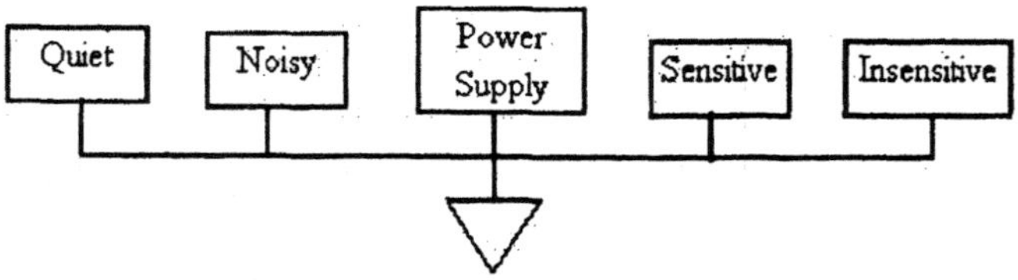

Figure 3.10. **Layout Rules for Sharing Returns**

Chapter 4

Basic Concepts Used in EMC

4.1 ANTENNAS

Many EMC issues result from energy that is transferred by radiation from a source. In order to understand this radiation of energy, it is useful to refer to some basic electromagnetic principles. One of these principles is the "isotropic point radiator" of energy. As this point source has zero radius and radiates equally well in all directions. This is shown in Figure 4.1.

Figure 4.1. **Isotropic Radiator**

Real energy sources that intentionally transfer energy by radiation are called "antennas" and have several key characteristics which differentiate them from isotropic radiators. The first is directivity, which is the direction of the maximum energy transfer. The second is gain, which relates to the shape of the energy transfer pattern.

If we look at the directivity of an antenna, it is essentially "the map of the gain" as shown in Figure 4.2. Gain refers to the ratio of any portion of the pattern to any other portion. In EMC work, another issue is antenna factor, which relates to the transfer function between energy and voltage at the terminals (discussed in detail in a later section).

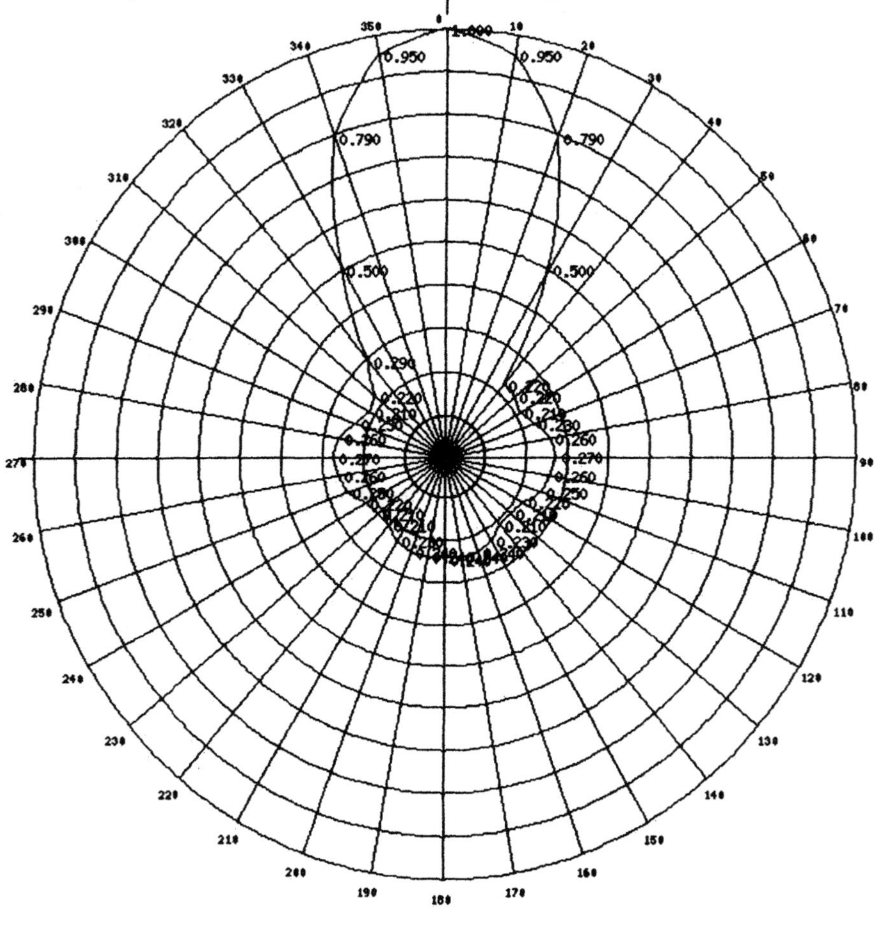

Figure 4.2. **Example of AM Broadcast Station Directional Antenna Pattern (Map of the Gain)**

We will now discuss basic antenna concepts and designs. This subject of the physics and mathematics behind antennas can be complicated and time-consuming. There are numerous references on antennas that the reader is encouraged to review for detailed understanding. Our intention in this text is to review basic antennas that may contribute to or create EMC problems.

Common Antenna Types

Two common types of antennas are "quarter wave" and "half wave" antennas. These names refer to the fact that their physical dimensions approximate a portion of the wavelength, which is determined from the speed of propagation and the frequency of intended operation (discussed previously). For example:

- A half-wave antenna used to receive a signal at 100 MHz would be approximately 1.5 m long
- An element of a quarter-wave antenna for the same frequency would be approximately 0.75 m long.

These antennas radiate with a maximum in directions 90 degrees from the axis of the elements. Consequently, these antennas are referred to as "omni-directional "antennas.

Another basic type of antenna is the "gain" antenna. This antenna differs from an omni-directional antenna in that this antenna both transmits and receives energy primarily from certain directions. (In some ways both a half-wave and a quarter-wave antenna exhibit some degree of directionality. While typically not defined as gain antennas, they do have characteristics that make them sensitive in certain directions, as shown in figures 4.3 and 4.4.)

In addition to directionality, another characteristic of these antennas is impedance at resonance (the radiation resistance). Radiation resistance means the effective resistance that the antenna exhibits when connected to a source. A half-wave antenna commonly used as a dipole antenna has a radiation resistance of approximately 73 ohms. A quarter-wave antenna, typically used with a counterpoise surface (generally called by many a "ground plane") has a radiation resistance of approximately 37 ohms.

Let's look in more detail at the dipole and quarter wave antennas. Dipole antennas are typically constructed horizontal to the ground and for communication purposes, are ideally located several wavelengths (at the frequency of operation) above the ground. Quarter-wavelength antennas are typically mounted with their main radiating element located vertically to the ground, and have one or more radials parallel with the ground. This is termed a ground-plane antenna because the radials approximate or are intended to approximate the earth itself. More correctly, the radials are the "counterpoise" for the antenna, and create an "image" element.

This antenna was developed to meet the need for an efficient, inexpensive base station antenna for use in communicating with mobile units. Most commonly seen with four equally spaced counterpoise rods, it turns out that work by Dr. George H. Brown of RCA showed that no more than two counterpoise rods are required.

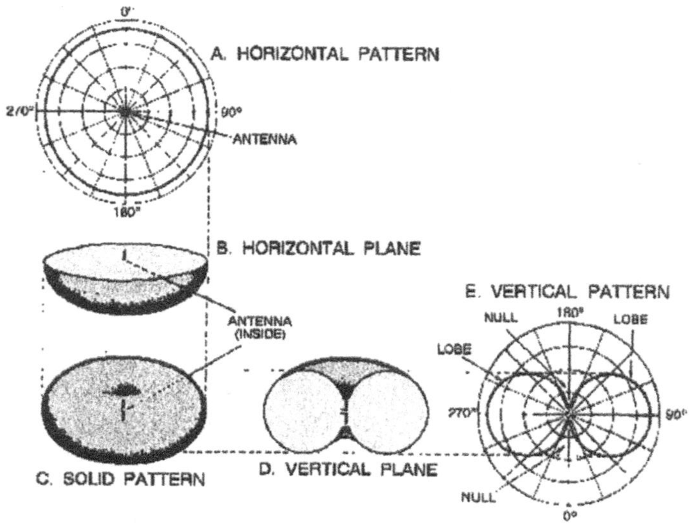

Figure 4.3. **Dipole Antenna Radiation Pattern**

Figure 4.3 illustrates both typical dipole horizontal and vertical antenna patterns. Figure 4.4 shows half-wave dipole and quarter-wave vertical antenna patterns.

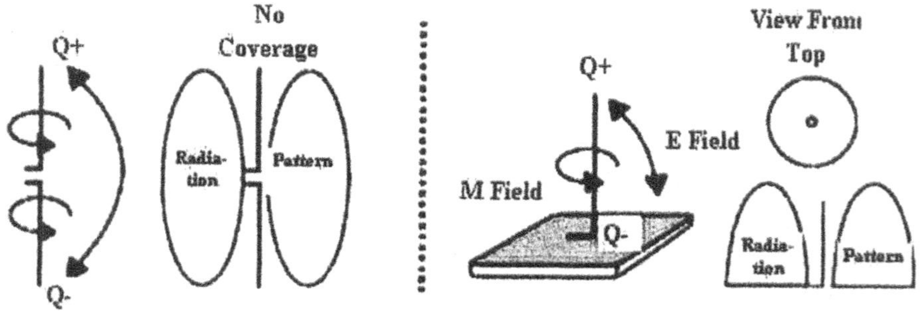

Figure 4.4. **Typical Antenna Patterns For Half Wave Dipole (Left) And Quarter Wave Vertical (Right)**

Figure 4.5 shows a typical ground-mounted vertical antenna installation.

Figure 4.5. **Vertical Antenna**

Figure 4.6 shows a typical dipole antenna. This antenna is mounted in a horizontal configuration, several wavelengths in height above ground.

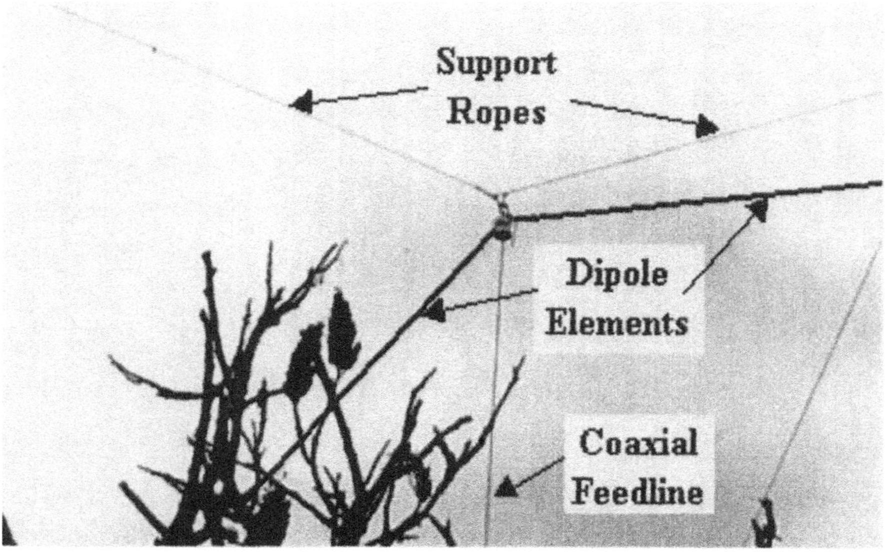

Figure 4.6. **Dipole Antenna**

4.2 OMNI-DIRECTIONAL ANTENNAS

4.2.1 Quarter-Wave Vertical

What are the dimensions of typical quarter-wave vertical antennas that are commonly used for mobile communications? The following are example calculations to use when determining the length of quarter-wave antennas. If you recall that the calculation for wavelength is equal to the speed of propagation (which in free space is 300 million meters per second) divided by the frequency in MHz, then the length of the vertical element of the quarter wave antenna would be the wavelength divided by four. Table 4.1 shows frequencies of common vertical antennas used for mobile communications and their approximate length:

Table 4.1. Frequency, Wavelength, and Quarter-Wavelength.

Frequency, MHz	Wavelength, meters	Quarter-Wavelength, meters, feet, or inches
27 (CB)	11	2.7 (9ft)
45 (Land Mobile)	6.7	1.7 (5.5 ft)
150 (Land Mobile)	2	0.5 (19 inches)
850 (Cellular Telephone)	0.35	0.09 (3.5 inches)

1 meter = 39.37 inches

4.2.2 Ground Plane

If we use the term ground plane for an antenna type, one meaning could be as shown in Figure 4.7, where we would have the vertical element over the reference or the "ground." As shown in Figure 4.9, the ground looks like the image of the vertical element or the counterpoise, which then resembles a one-half-wave dipole. The difficulty when referring to these types of antennas can be seen in this example; if we have a vertical element on an aircraft in flight as in Figure 4.8, where is the ground plane?

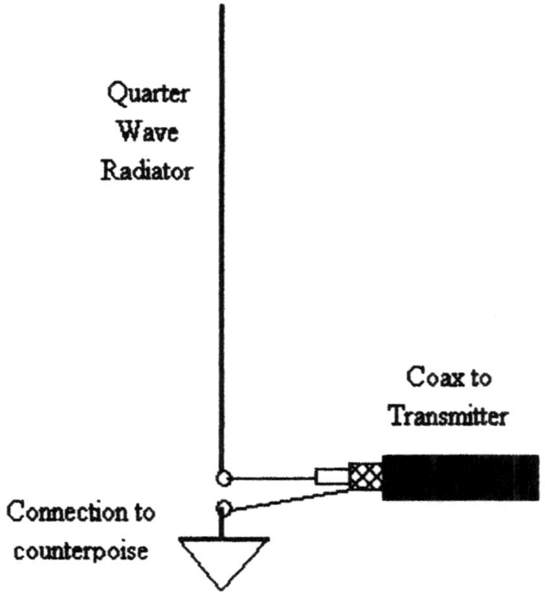

Figure 4.7**. Basic Representation Of A Simple Vertical Antenna**

Figure 4.8. **Ground "Plane" Antenna or "Grounded Plane Antenna"**

A quarter-wave perpendicular to a reflecting plane is electrically the same as a half-wave dipole.

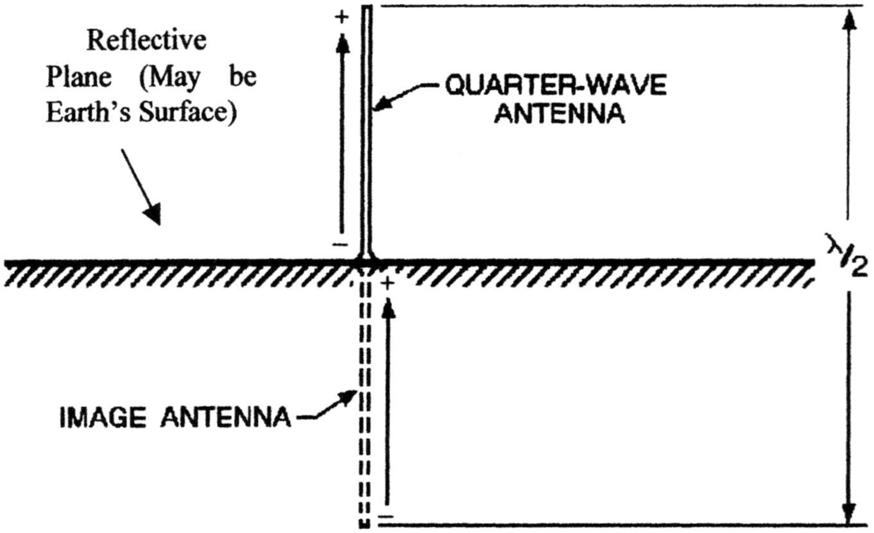

***Figure 4.9.* Ground Plane Antenna Operates Because Of The Creation Of An "Image" Antenna**

4.2.3 Other Antenna Types

4.1.3.1 Antenna Arrays

Another way of obtaining antenna gain is the method used to provide "directional" capabilities to fixed broadcast stations. This is accomplished by using individual antennas in an "array" configuration. Some AM broadcast stations in the United States are required to provide a directional broadcast pattern in the evening to prevent interference to other stations. Typically accomplished by feeding different antennas in an array, this is an example of how radiation from antennas can cancel each other and form a directional pattern. We will discuss this in Chapter 8 when we cover differential and common mode radiation.

4.2.3.2 Unanticipated Antennas

In addition to intentionally creating antennas, connecting conductors to components creates a system that did not exist when considering only the components. The contribution of the conductor results in increased efficiency of energy transfer and behaves like an antenna at lower frequencies than would be possible with just the component itself, which is a smaller size than the combination of the conductor and the component. Empirical data suggest that a conductor longer than 10 percent of a particular wavelength starts to become an efficient radiator. For example, a printed circuit board trace with a length as short as approximately 0.15m (6 inches) could be an efficient radiator of the system emissions at approximately 200 MHz! The ten percent rule is reasonable, since a quarter wave antenna is an excellent radiator.

The significance to EMC is that system radiation can be confusing when evaluating component level test results that appear to conflict with the component dimensions. (See Figure 4.10.) If the dimensions of the component are expressed as D_0, and emissions from the component are plotted as energy transfer versus wavelength of the energy, this is shown to the far right of the graph. If the length of the conductor is expressed as D_1, and the energy transfer as a function of wavelength for the conductor is also plotted, this would move to the left slightly. Now, however, if the energy transfer for the component and the conductor (D_0 +D_1) is plotted, this would be shown in the curve further left in the Figure. This is again using the estimate that energy transfer increases significantly as the length of the conductor becomes greater than 10 percent of the wavelength. Compare this to our standard antenna that we previously discussed. If we use the one-quarter-wavelength antenna to compare its dimensions of this 10 percent statement, then this is true since the one-quarter-wavelength antenna would be 25 percent of the wavelength, which is significantly more than 10 percent of the wavelength.

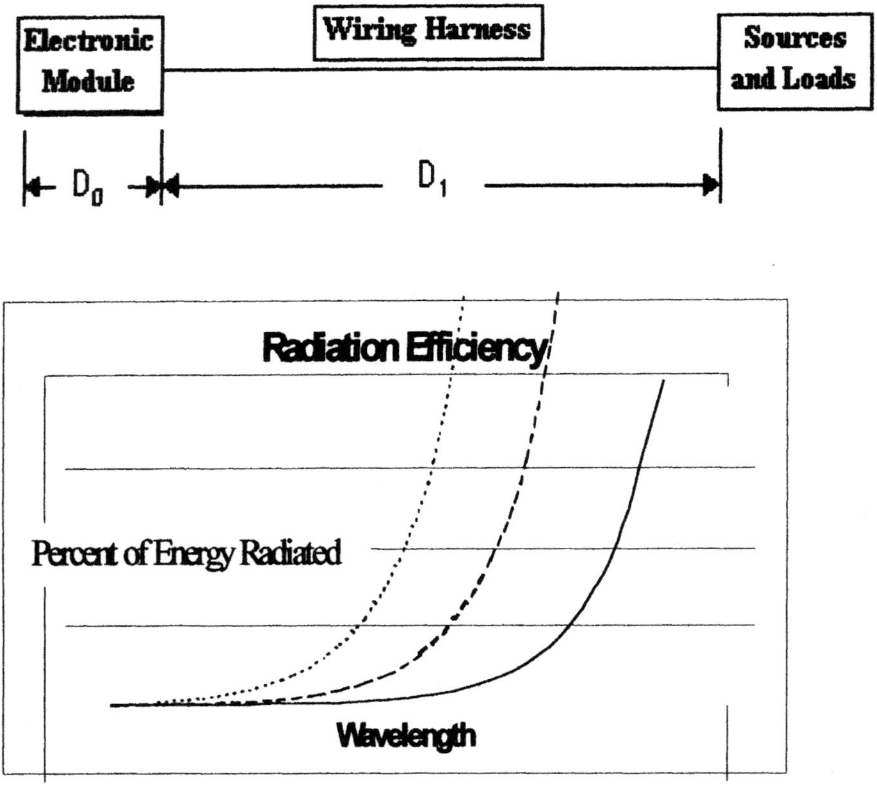

Figure 4.10 **Role Of Conductor Length In Energy Transfer**

Legend:

Length D0+D1 ....

Length D1 - - - - - - - - - - - - -

Length D0 ——————————

4.2.3.3 Reduced Size Antennas

Another type of antenna can be created by incorporating a series inductive element in a conductor that is physically shorter than the electrical length of a quarter-wavelength antenna. By use of this inductive element, the antenna can be physically shorter then one-quarter of the wavelength at the frequency of use, yet be at a quarter wavelength electrically. This principle is used on vehicles that utilize HF frequencies for communications. Without the use of the inductive element, the antennas would be very long and impractical on a vehicle. The reason that this antenna can utilize an inductive element is that the addition of the inductive element makes the antenna appear "electrically" one-quarter of the wavelength, and therefore is closely matched to the transmitter impedance of 50 Ohms. This technique is also used in portable electronic devices in order to reduce the physical size of the device, such as a cellular phone, and still have an efficient radiator of energy (Figure 4.11). Unfortunately, from the automotive EMC standpoint, what this can also mean is that unintended inductance in cables or wiring can result in creating efficient antennas when one was not desired. For those readers that are interested in the amount of inductance required as a function of length of a conductor, references are included in this chapter.

There are some configurations of vertical antennas that are also gain antennas. What these antennas essentially do is to modify the effective radiation pattern from the standard quarter-wave antenna (shown in figures 4.12 and 4.13). This radiation pattern is more concentrated along the lower angles, near the horizontal plane, effectively "increasing the range" of the antenna. The energy that escapes to the sky in a mobile communications system is energy that does not contribute to the signal. The approximate pattern around the gain antenna in the vertical configuration is shown in Figure 4.12. Top views of both these antennas are also shown. In the Figure, circular lines around the center of the antenna indicate regions of constant field strength. What we will see is that the field strength from the lower angles of radiation of the gain antenna is higher at a further radius out from the antenna than for the quarter-wave antenna. Another common type of gain antenna used in both communication systems (or radio and television reception) as well as EMC work is titled the Yagi. These antennas are sometimes referred to as "Fishbone" antennas. In general, these types of gain antennas consist of a number of elements perpendicular to the antenna boom, used to support the elements. At the rear of the boom are elements that are relatively long with respect to the elements at the front. This indicates that the energy would be received most effectively in the direction of the shorter elements. These antennas can be very efficient and have high values of gain, sometimes from 10 to 20 dB! This means that, if we have 10 dB of gain, this is equivalent to an increase of 10 times in the signal power. Likewise, 20 dB is a 100 times increase of the signal power (if we go back to the equations). Gain (dB) = 10 Log (P_2/P_1), where

P_1 =Effective Radiated Power of Reference Antenna
P_2 = Effective Radiated Power of Gain Antenna

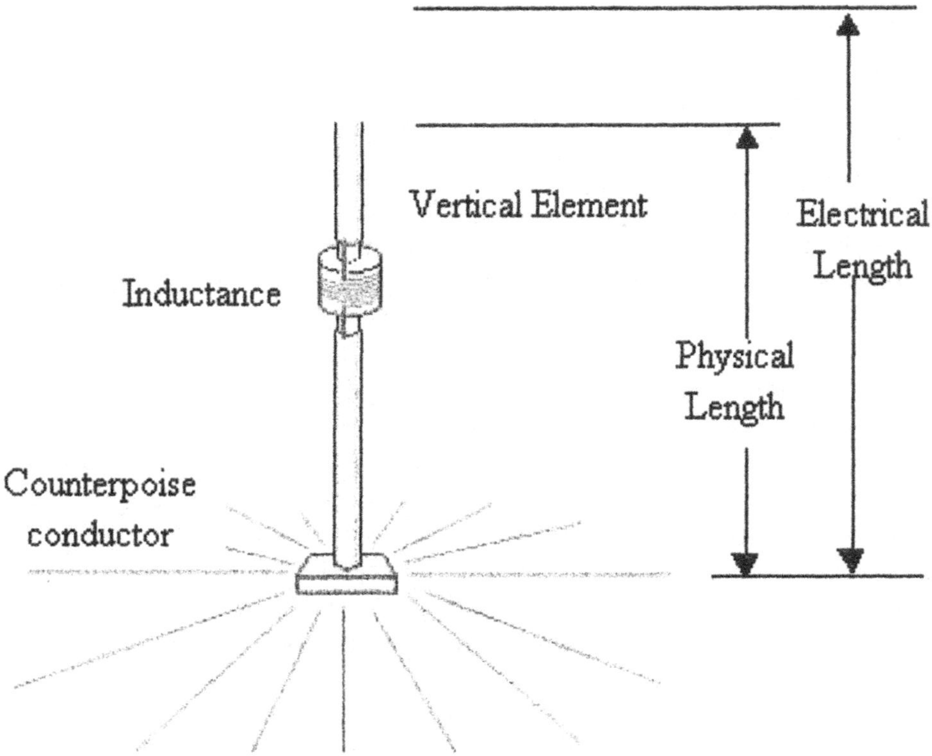

Figure4.11. **Center Loaded Vertical Antenna**

4.2.3.4 Gain Antennas

Figures 4.14 and 4.15 show examples of specific gain antennas, referred to as Horn and Log Periodic, respectively. As discussed earlier, both these are directional antennas and have gain.

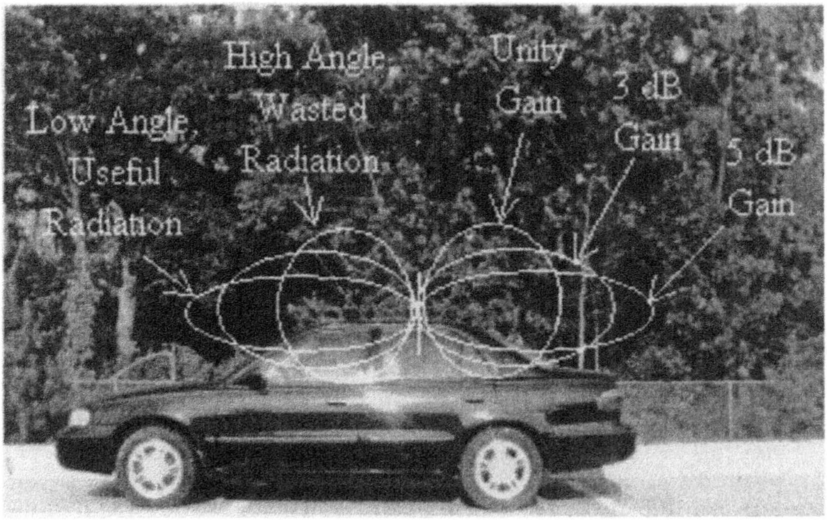

Figure 4.12. **Side View of Vertical Monopole Radiation Pattern**

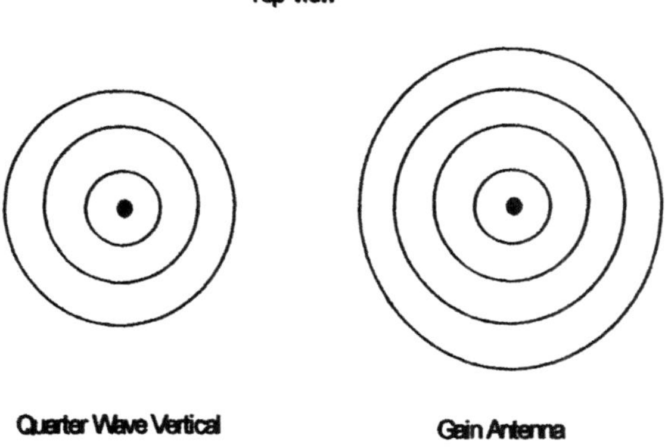

Figure 4.13 **Top View Of Monopole Radiation Pattern**

Figure 4..14. **Horn Antenna**

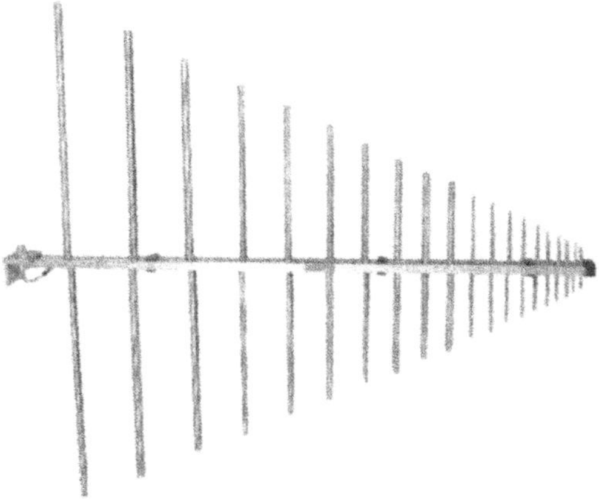

Figure 4.15. **Yagi Antenna**

4.3 OTHER COMPONENTS IN EMC

We will now discuss basics of other components used in EMC work. It is important to understand the physics of each of these components, as they comprise a large portion of the intended and unintended current paths in EMC. The remainder of this chapter will discuss:

- Inductors
- Capacitors
- Resistors
- Cables and Transmission Lines
- Shields

4.3.1 Inductance

Many people use the term inductance without really understanding what it means. Inductance is a derived unit and is the property of a current loop to generate a voltage in proportion to the rate change of current through that loop. The basic unit of inductance is the "Henry".

> 1 Henry is defined when one Volt is generated by a current change of 1 Amp per second.

The relationship between Inductance, Current rate of change, and Voltage is defined in the equation (utilizing calculus notation):

$V = - L\, dI / dT$,

Where V = voltage developed
dI = the change of the current (in amps)
dT = the time it takes for the current to change from its maximum to its minimum

For most practical purposes, rather than requiring knowledge of the actual equation for the derivative, it is acceptable to utilize a linear determination of the rate of change, which would be represented by the term "delta". This equation is important to understand when determining the impact that inductive devices have upon a circuit's operation.

What are inductors from a physical standpoint, and what does this mean for wiring for automotive EMC systems and components?

In theory, the inductance is equal to the sum of the internal inductance L_{ii}

and external inductance L_e of the wire. In reality, the internal inductance is at most one percent of the external inductance, and can be neglected. Therefore, the PRIMARY contributor to inductance is conductor loop geometry. This can be seen in the following equation for a single turn loop in air.

$$L\ (\mu H) = (A/100)*(7.353\ LOG\ (16A/D) - 6.386)$$

Where A = Radius of loop in inches
D = Diameter of wire in inches

It is likewise important to point out another aspect of inductance that is a misunderstood term in the electrical electronic work, both between the neophyte and the experienced engineer. There are two terms that are used in describing inductance. The first one is titled "self-inductance" and the other is titled "mutual inductance". All that self-inductance refers to is the property of an inductor to create a voltage opposing the rate of change of the current passing through it. As was stated before, it is important to understand that there must be a complete current path. This means there must be a loop of current in order for self-inductance to occur.

Mutual inductance means is the ability of one inductor to induce a current into a second inductor. Understanding the physics is important because this means that the inductance of a straight piece of wire is zero! This can be demonstrated through a very simple experiment, consisting of taking a length of wire and arranging it into a loop enclosing some area and measuring the inductance. The next step would be to take a much longer length (let's say two times) and arranging that in another loop and measuring inductance. If the inductance were a property due to lengths of straight pieces of wire, then it would have to mean that the inductance would be exactly twice that for the longer piece of wire than for the shorter piece of wire. Experiments can be very simply conducted showing that the inductance for the second piece of wire, which is much longer, can actually be less than that for the first piece of wire. The equations for inductance of loop can also be used, which will show that it is possible to have less inductance with longer wire. This is because the inductance is related to the area enclosed by the loop, not the length of the wire. Another term for inductors which you may have heard or used before, is a "choke". What this implies is that the inductor is choking the current flow by increasing the impedance.

4.3.2 Inductance of "Large" Wire Loops

In his book *PCB Design for Real – World EMI*, Dr. Bruce Armchambault has a comprehensive discussion of how to determine inductance of several types of loops, some of which are shown in figure 4.16.

A computerized calculation method for various loop geometries has been developed by the University of Missouri – Rolla. They have an "Inductance Calculator" that can be accessed at their web site:

http://emcsun.ece.umr.edu/new-induct/

The authors of this text have used these calculations and the values obtained compare well to actual measurements made on various automotive wiring configurations.

4.3.3 Capacitance

Another fundamental component is the capacitor. You may recall that the capacitor consists of two conductors that are separated by a dielectric, as shown in Figure 4.17.

4.4 Ideal and Actual Components

When working on EMC issues it is important to remember that we are dealing with actual components, unlike textbook theoretical components. For example, inductors have only inductance, capacitors have only capacitance, resistors have only resistance, etc.

A key question to ask at this point is how many products are built from only theoretical components? The answer is simple-none! So another issue in EMC work to comprehend is that there is a difference between ideal components and actual components. The difference between an ideal and an actual capacitor is shown in Figure 4.18, and the ideal vs. actual inductor in Figure 4.19. The ideal inductor has only inductance, and actually the real inductor has parallel parasitic capacitance, as shown. The ideal capacitor has only capacitance, while the real capacitor has series inductance as well.

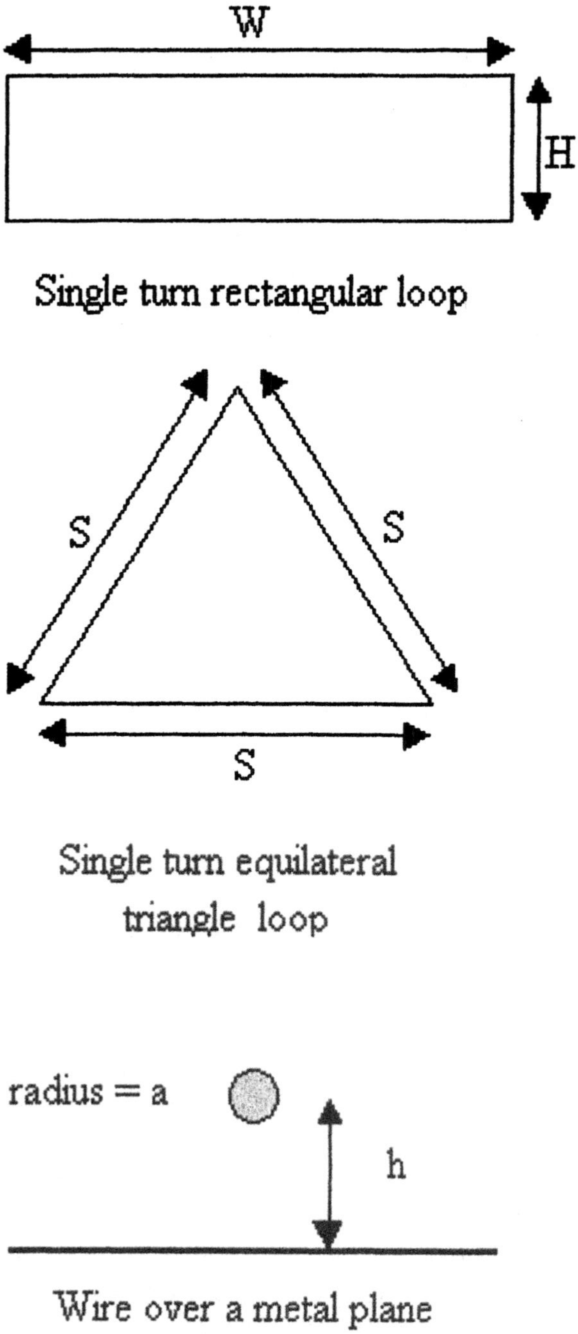

Figure 4.16 **Several Types of Wire Loops**
Courtesy of Bruce R. Archambeault. Used with permission

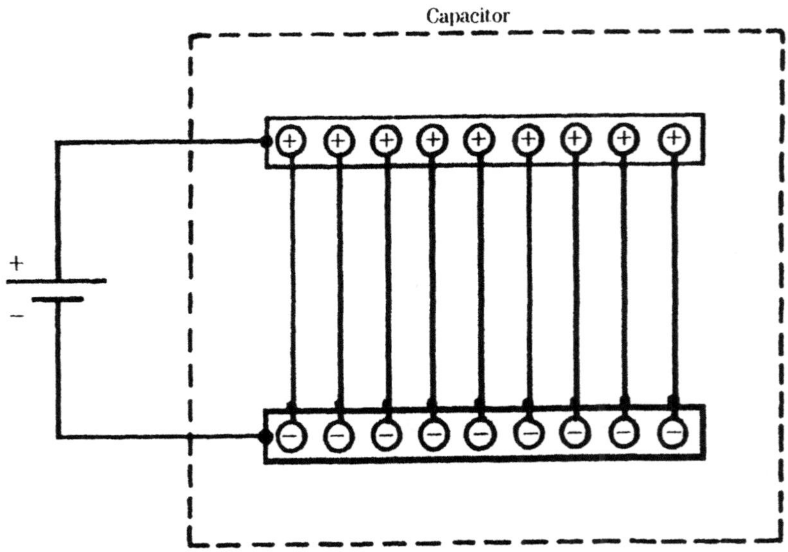

Figure 4.17. **Capacitor Showing Electric Field Between The Plates**

A refresher about capacitance follows:

- Capacitance is Proportional to Area of plates, and Dielectric Constant.
- Capacitance is inversely proportional to the distance between the plates.

Figure 4.18 shows the following for a real capacitor:

- At low frequencies, the capacitance is dominant.
- At high frequencies, the inductance is dominant.

Because of self-resonance effects, certain types of capacitors are preferred for use at specific frequencies, as shown in Table 4.2.

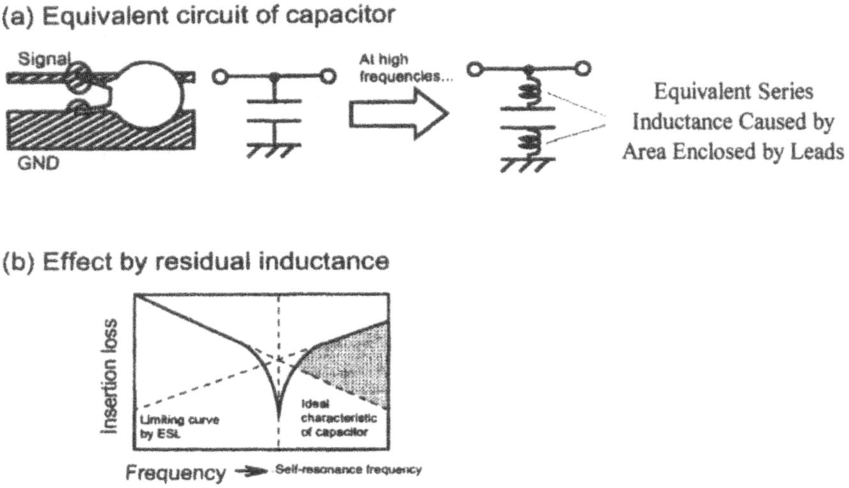

Figure 4.18. **Ideal Vs Actual Capacitor**

Table 4. 2. **Optimum Frequency Range For Different Types Of Capacitors**

Type	Frequency Range
Aluminum Electrolytic	1 Hz – 10 kHz
Tantalum Electrolytic	1 Hz – 10 kHz
Paper or Mylar	100 Hz – 5 MHz
Ceramic	1 kHz – 100 MHz
Plastic Film	1 kHz – 9 GHz
Mica, glass, or ceramic	5 kHz – 10 GHz

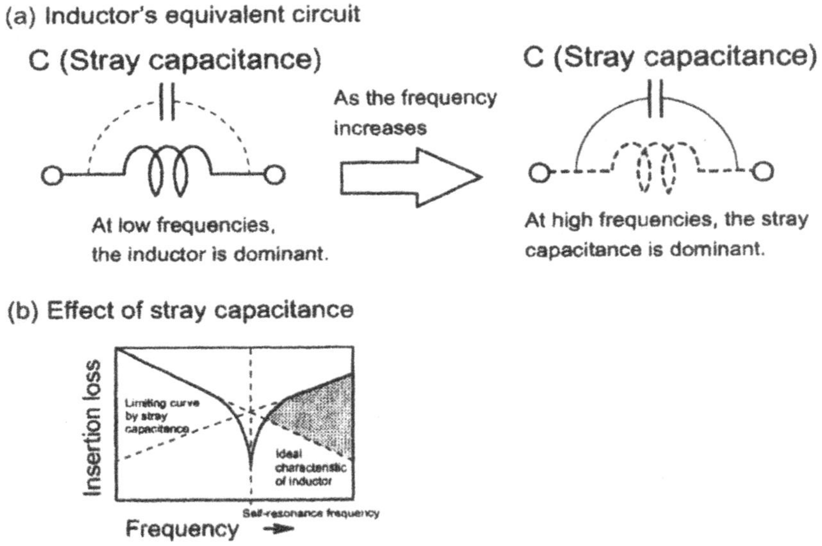

Figure 4.19 **Ideal Vs Actual Inductor**

In figures 4.20 and 4.21 respectively, the impedance vs. frequency of the ideal capacitor and the ideal inductor are shown. As the frequency increases, the magnitude of the impedance (abbreviated as the capital letter Z) for the ideal inductor increases in direct proportion to increasing frequency. The impedance of the ideal capacitor decreases due to the inverse relationship between capacitance and frequency. If we consider the inductance as represented by a series inductor and a parallel capacitance, then at some frequency the impedance of the component will decrease as frequency increases. If we consider the capacitor as being represented by a capacitance and a series inductance, then at some frequency the impedance will increase as a function of frequency. If people are unaware of the frequency-dependent impedance of actual components, this can cause problems in what would otherwise seem to be a good circuit design or implementation. For example, in designs including operational amplifiers, it is very important to understand the key impedance characteristic of the capacitors that are used in the circuit. This phenomenon is referred to as the self-resonant point of the device, and is typically abbreviated as fr. The self-resonant point can be calculated by using the formulas for resonant circuits. The main issue is understanding the values of the parasitic elements and the intended device.

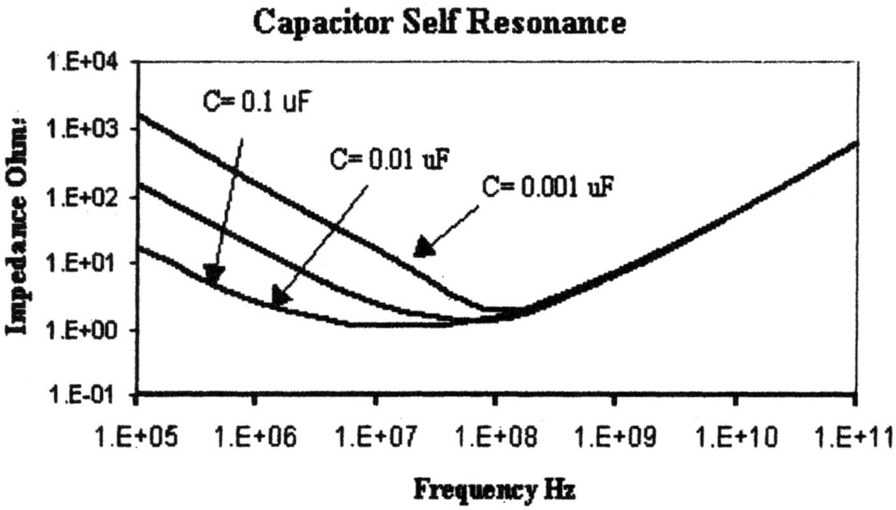

Figure 4.20 **Capacitor Self Resonance**

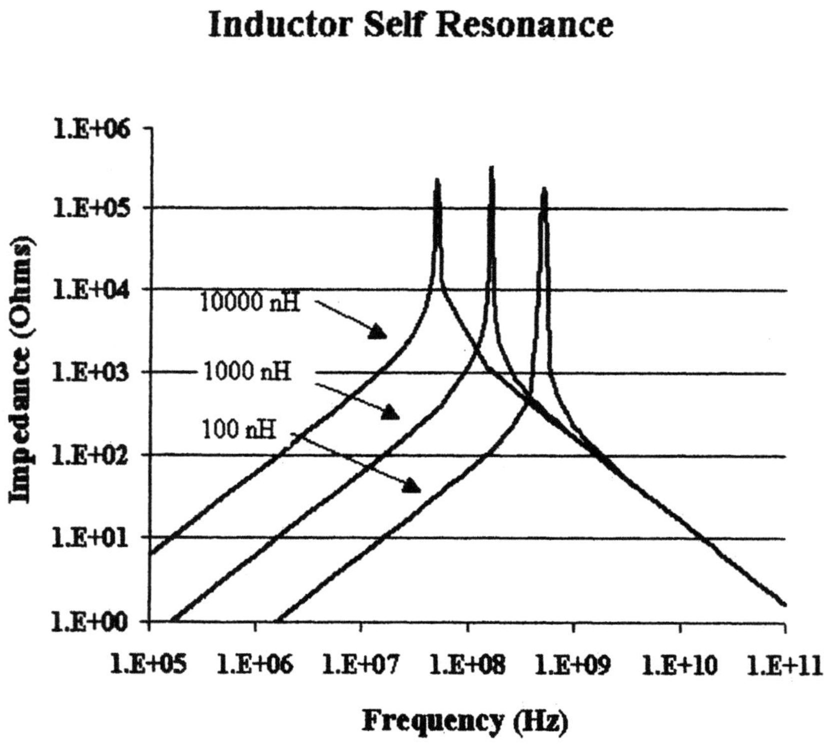

Figure 4.21 **Inductor Self-Resonance**

Different types of inductors have different characteristics, as shown in Table 4.3. Leakage flux should be minimized to avoid coupling to nearby circuits.

Table. 4.3. **Inductor Characteristics**

Inductor Type	Leakage Flux
Open Core	High
Closed Core	Low
Toroid	Very Low

A third fundamental component is the resistor. Resistor types are summarized in Table 4.4. Depending on the type of construction, resistors may also exhibit frequency sensitivity and self-resonance. Check the manufacturer's specification sheets to ensure that the impedance of the resistor is not affected by the signal frequencies.

Table 4.4 **Resistor Characteristics**

Resistor Type	Cost	Frequency Sensitivity	Current Handling Capability
Carbon	Low	Low	Low
Wire Wound	High	Inductive	High
Thin Film	Medium	Moderately Inductive	Low
Chip	Medium	Moderately Inductive	Very Low

4.5 TRANSMISSION LINES

A transmission line is used to transfer energy from the antenna to a transmitter. See Figure 4.22.

There are several types of transmission lines and we will discuss the most common ones here. They all have similar characteristics and functions; however, the construction can be very different and each type can have very different requirements. Figures 4.23 and 4.24 show the two primary types of transmission lines, "open wire line" and "coaxial line". There are important differences between these two types of transmission lines. The first one that can be seen is that the open wire line in some ways resembles a ladder, with the conductors located opposite from one another and the insulator between them. This transmission line has some positive characteristics as well as some disadvantages. One advantage of this type of transmission line is that connections to it can be made very easily; no special connectors are required. Another advantage is that it is fairly inexpensive, and exhibits low loss. The disadvantages are that, because of its construction, it provides very little shielding from the external noise and does not have the ability to prevent energy from being radiated from the transmission line. Coaxial cable compensates for some of these disadvantages; however, it is typically more expensive and does require special connections usually consisting of cylindrical connectors that must be fastened to the coax so it can be connected to the mating connector. Another disadvantage with coax is higher loss, resulting in less of the power being transferred from the antenna to the receiver or from the transmitter to the antenna. Also, the wave propagation speed is less than the propagation in free space, which means that the wavelength is different within the cable than in free space. The propagation velocity in the cable is less than in free space; therefore, the wavelength in the cable is somewhat shorter.

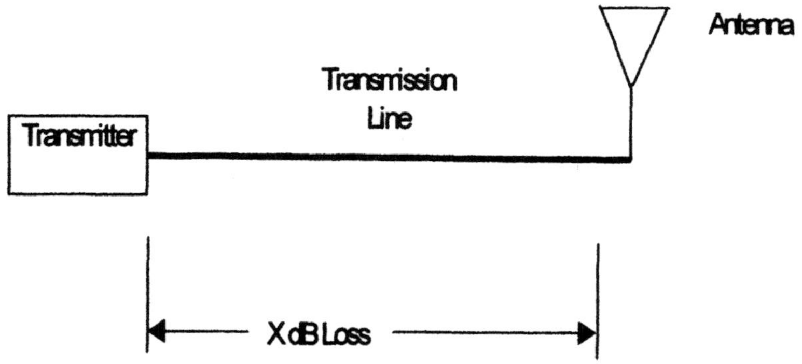

Figure.4.22. **A Transmission Line Connects The Transmitter To The Antenna**

4.5.1 Characteristics of Commonly Used Transmission Lines"

Table 4.5 shows the characteristics of commonly used transmission lines.

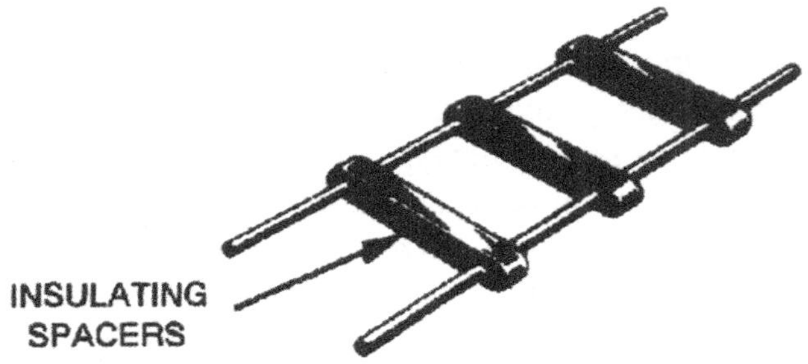

Figure 4.23. **Open Wire Transmission Line**

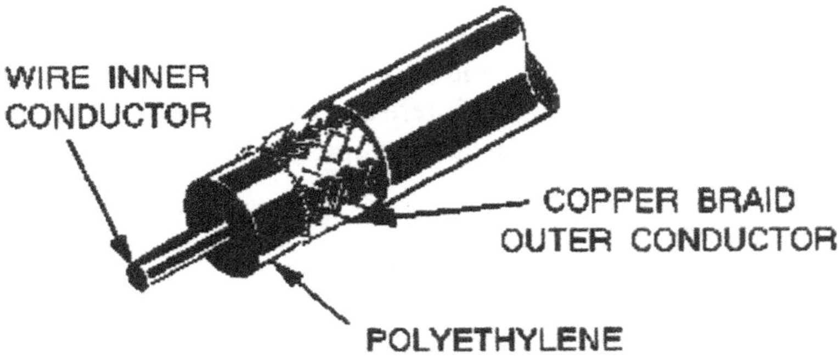

Figure 4.24. **Coaxial Transmission Line**

There are many types of coaxial cables, many with similar impedance, although they may vary in terms of their maximum power handling capability. The maximum power handling capability is a function of the dielectric material used for the insulation, and the amount of voltage that the dielectric can withstand. Another key item is the outside diameter (OD) of the cable. It can be seen that, in general, as the OD increases, so does the maximum voltage rating. There are several types of dielectric material that are used, including air, in the cable. The other type of cable shown is the parallel line or the "open wire" line. The impedance of the cable is a function of the distributed inductance and capacitance. It can be seen that, as the conductors are moved apart, the capacitance decreases, and the maximum operating voltage that can be applied increases, and the impedance also increases.

4.5.2 Goal of transmission line

The use of a transmission line is fundamentally the method that we employ to transfer energy from a transmitter (or RF source/amplifier) to an antenna, transducer, or some other type of RF load. See Figure 4.25. There is another important item of concern, and that is the impedance match from the transmission line to the load. To refresh the concept of maximum power transfer, a simple analogy is seen from dc circuits, that of matching the load resistor to the source resistance. This is still true with transmission lines and RF energy, except resistance is replaced by impedance. Maximum transfer takes place when Zload = Zcable. This results in zero reflected power. From a practical matter, there will be some amount of reflected energy, since no real components are perfect. The formula for the "reflection coefficient" Γ is shown below. There is also a more common term that is used, that of voltage standing wave ratio, and that is related to reflection coefficient. From a practical application, high reflection coefficients can cause damage to the actual test equipment and/or, at the minimum, produce misleading results.

Reflection coefficient $\Gamma = (Z_A - Z'_0)/(Z_A + Z_0)$
Z_A is the load impedance,
Z_0 is the transmission line characteristic impedance,
Z'_0 is the complex conjugate of Z_0
$SWR = (1+|\Gamma|)/(1-|\Gamma|)$

4.5.3 Transmission line capacitance

The transmission line model includes distributed capacitance and inductance. This means that for a given length of transmission line, there will be essentially a parallel

Table 4.5 **Feed Line Loss for Commonly Used Feed Lines (dB/100 Feet)**

Frequency	RG-58	RG-8X	RG-8A, RG-213	RG-8 Foam	9913 & Equiv	Half Inch 50Ω Hardline	Open Wire
1.8	0.5	0.4	0.3	0.2	0.2	0	0
3.5	0.7	0.5	0.4	0.3	0.2	0.1	0
7	1.1	0.7	0.5	0.4	0.3	0.2	0
10	1.4	0.9	0.6	0.5	0.4	0.2	0
14	1.7	1.1	0.8	0.6	0.5	0.3	0
18	2.0	1.2	0.9	0.7	0.6	0.3	0.1
21	2.2	1.3	1.0	0.7	0.6	0.3	0.1
24	2.4	1.4	1.1	0.8	0.6	0.3	0.2
28	2.5	1.5	1.3	0.9	0.7	0.4	0.2
50	3.5	2.1	1.7	1.2	0.9	0.5	0.3
150	6.5	3.6	3.0	2.0	1.6	1.0	0.7
220	8.4	4.6	4.0	2.6	2.0	1.3	
450	12	6.5	5.8	3.6	2.8	1.9	
900	19	9.6	9.0	5.4	4.0	3.0	
1200	23	12	11	6.4	4.6	3.7	
2300		15	15	8.8	6.4	5.2	

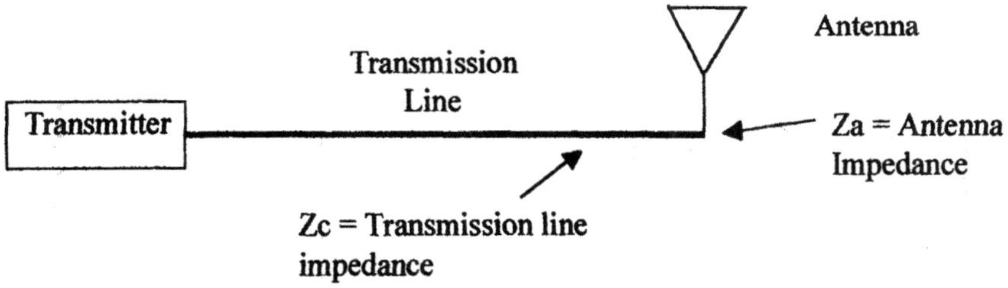

Figure 4.25 **Transmitter Feeding Antenna Through a Transmission Line**

capacitance that is a function of the length of the cable. This can be verified by the simple experiment of connecting a transmission line to capacitance-measuring equipment. One of the easiest ways to do this is to use a section of coaxial cable. Most coax cable has a capacitance of 10's of picofarads per foot. It is seen that, if the cable is connected, and a certain capacitance is measured, if the cable is doubled in length, the capacitance will be doubled.

4.5.4 Transmission line impedance

It seems like the subject of transmission lines for RF is one of the most misunderstood concepts in engineering. We will discuss the characteristics of transmission lines and relate those characteristics to actual physical attributes, and then study examples of transmission line geometry.

A transmission line can be modeled as shown in Figure 4.26, where the line is expressed as a continuous sequence of similar lumped elements that relate to the per-unit length characteristics. This results in a couple of parameters; the first is that of "characteristic impedance" of the line. It is important to understand that this impedance is NOT a resistance. It is a complex relationship between the reactance of the distributed elements and the frequency of the RF signal. Another item is that of the propagation speed of the signal as it moves along the line. Remember that in a vacuum, the propagation is at the speed of light. In a transmission line, the speed is only a portion of the speed of light, with the result that the effective wavelength of the signal is reduced only in that cable. The typical propagation speed (called the velocity factor with respect to the speed of light) is about 60 to 80 %. The characteristic impedance is normally about 50 to 100 ohms for coaxial cable, and is about 300 ohms for open-wire line, or "twin lead'.

The value of the characteristic impedance is as follows:

- Zo = square root of L/C.
- Where L is the distributed inductance
- Where C is the distributed capacitance

It can then be seen that the impedance varies as a function of the relationship of the conductors to one another. For example, if the conductors are moved farther apart from each other (with all else being the same) the capacitance between the conductors decreases. This results in the Zo increasing. See Table 4.4 for characteristics of transmission lines, where the open wire type has a higher Zo.

Other characteristics of open versus coaxial cable:

- Open wire is a low loss cable that is easy to connect to, sometimes having common spade terminals.

- Coaxial cable requires special connectors, which can be difficult (and expensive) to install. These connectors are needed, however, to maintain the correct geometry to ensure Zo consistency.

4.5.5 How to install a PL 259 connector

Since coaxial cables are used quite frequently in EMC and RF work, a review of how to install on the most common type of connector is appropriate. This connector has been used for many years and is called a "PL-259" or "UHF_2" connector. Figure 4.27 shows how to install this type of connector. As with any type of coaxial connector, it is important to maintain the integrity of the inner conductor and the outer shield, to maintain the dimensional stability, in order to maintain the impedance. There are two key parts of installing coaxial connectors: the first one is to remove the outer insulation, and the second one is to prepare the center conductor. It is important to be precise when installing the connector onto the cable, so that the shield or the center conductor is not damaged in any way. Understanding exact dimensions of the required length for the shield and the center conductor all are also important to provide enough connection for this solder to be properly applied. By following the example shown here, a good connection can be made.

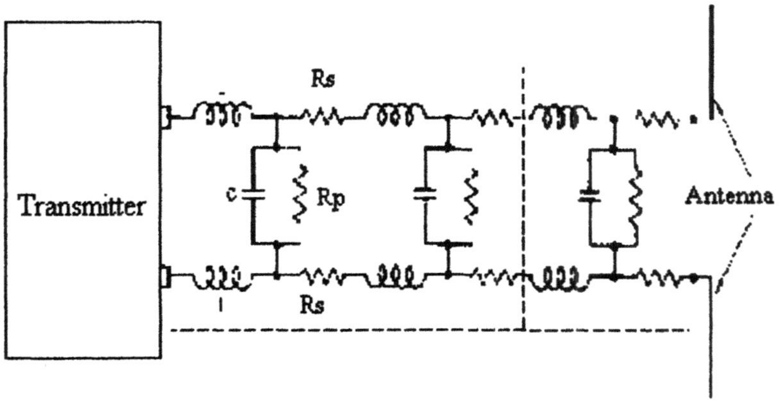

Figure 4.26 **Transmission Line Model**

4.5.6 Coax Cable Sample

Figure 4.28 shows typical coaxial cables and open-wire lines. In the photos, the center conductor of the cable is shown, with the insulation around the conductor. The outer conductor (called the shield) is shown. It looks like a braid of wiring around the center conductor's insulation. The figure also includes coax cables of various sizes and impedance.

The wire in the other photos shows various types of open wire or twin lead cables. It can be seen that these are very simple in construction, with two conductors on each side of an insulator. An additional feature of coaxial cable is that it does provide electric field shielding from external sources as well as radiation from the cable itself. The reason for this is shown in Figure 4.29. Sometimes the rationale for the shielding effectiveness of coax cable seems to contradict what is believed to be true for how RF energy travels on a wire. It has been shown that RF travels along the outside of a conductor, and this is true for the energy along the center conductor of the coax. If we say that the RF travels along the outside of the conductor on the shield, this would seem to indicate that the RF would be along the outside of the coax, and thus would seem to NOT have the shielding properties. In fact, the current does travel along the outside of the center conductor, and because of that, the return current for the energy on the center conductor flows along the INSIDE surface of the "shield" because the charge is similar to a capacitor, where there is an electric field between the conductors, and is actually in the dielectric material.

From this discussion, it is easy to understand why open wire lines are not shielded from external (or can radiate) energy, since the signal and the signal return travel along the wires and are exposed to the environment. The terms balanced and un-balanced have also been used to describe transmission lines. This relates to the coupling between the conductor and the other conductor, as well as the external environments, as is shown in Figure 4.30.

("UHF" Does not mean this connector is a good impedance match at high frequencies)

If we represent the coupling of the conductors to each other and the external environment, it can be seen that the open wire line conductors have each the same amount of capacitance coupling to each other and the environment. In the case of coaxial cable, the center conductor is coupled to the shield, and the shield is coupled to both the center conductor as well as the external environment, resulting in an unbalanced condition.

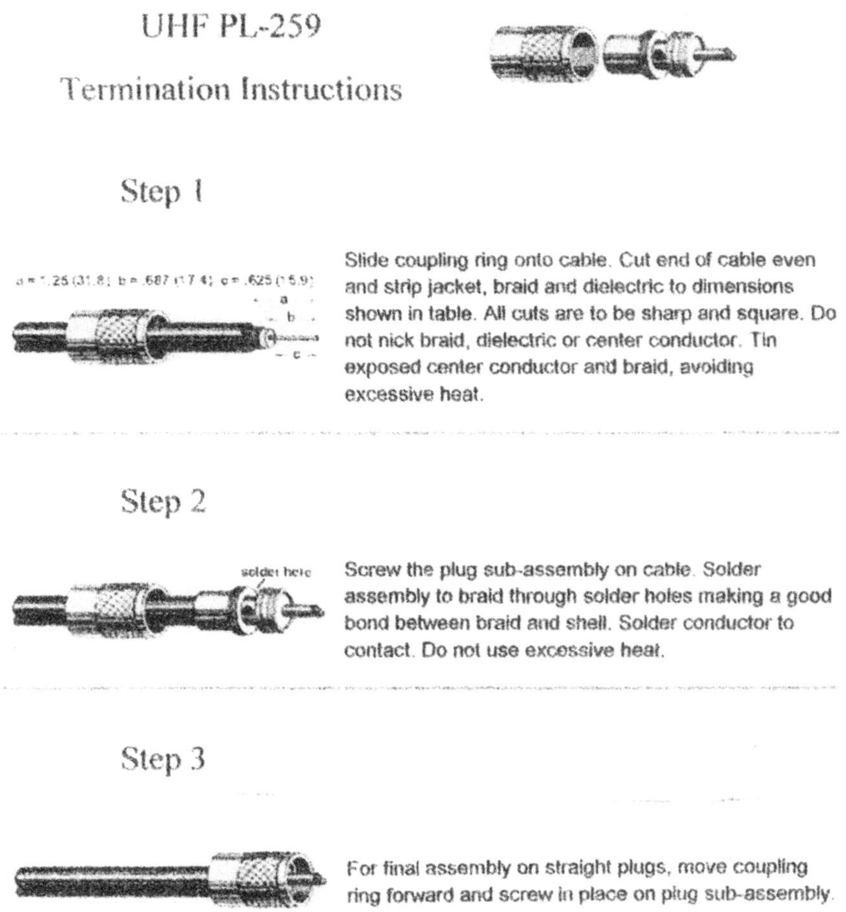

Figure 4.27. **Installing A "UHF" Connector On Coaxial Cable**

In summary, we've discussed the various types of coax cable as RF transmission lines, and the tables of information that exist to assist in the selection of cables. We've also looked at the dimension of the cables, and seen that the connections to coax are more complicated than with open wire lines.

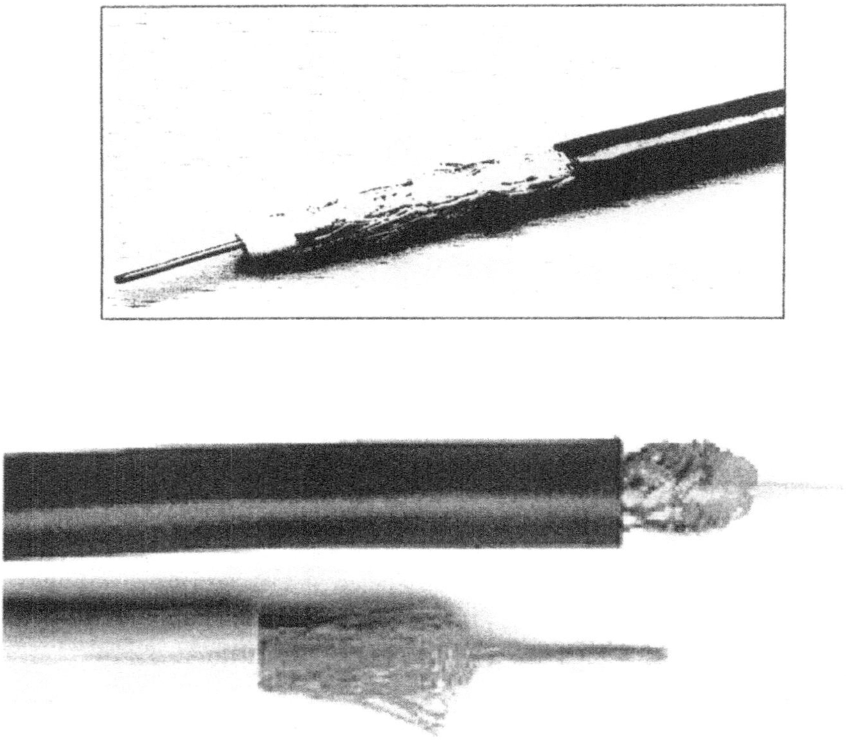

***Figure 4..28.* Various Types Of Coaxial Cable**

What is also important to understand is that the selection of cable depends upon several factors, including knowledge of the loss factors of each type of cable, physical dimensions, and impedance (typically for EMC work the cable impedance is 50 or 75 ohms).

If it turns out that that the immunity problems are due to electric fields being picked up by the wiring and conducted into the device, there are some techniques that can be used to provide wire shielding. There is much confusion as to how to shield for electric and magnetic fields. We will discuss the fundamentals of this shielding and how the techniques for H and E field shielding are different.

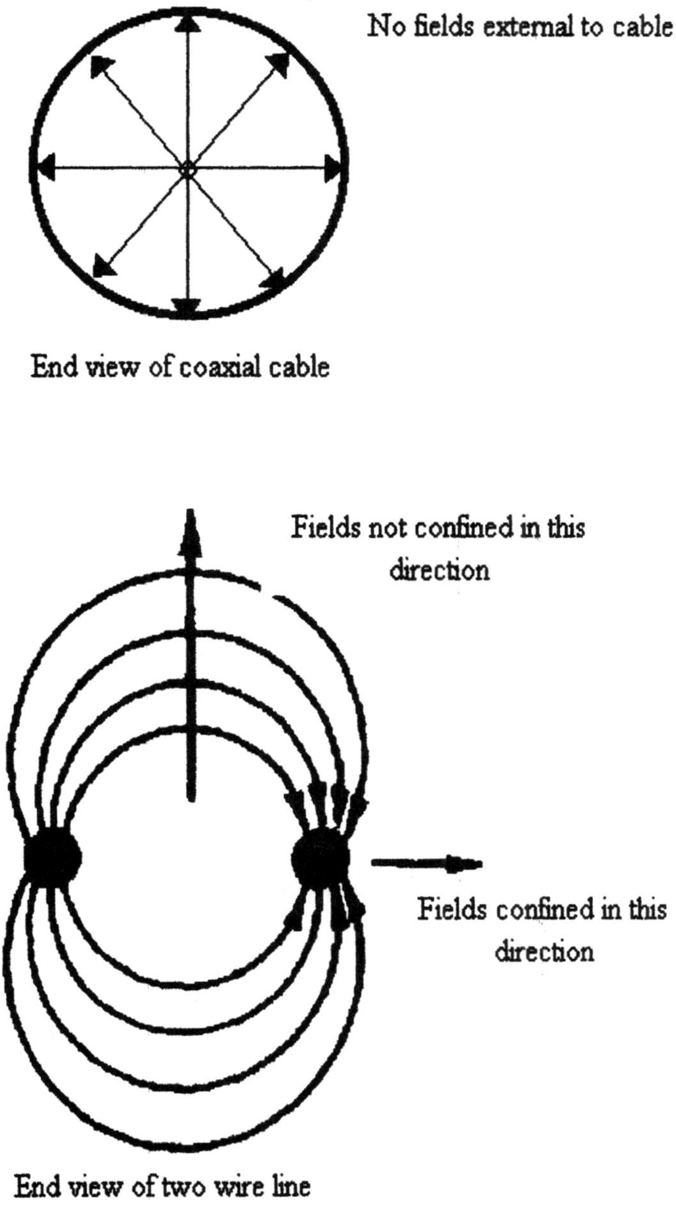

***Figure 4.29*. Coaxial Cable Radiation Vs Open Wire Line Radiation**

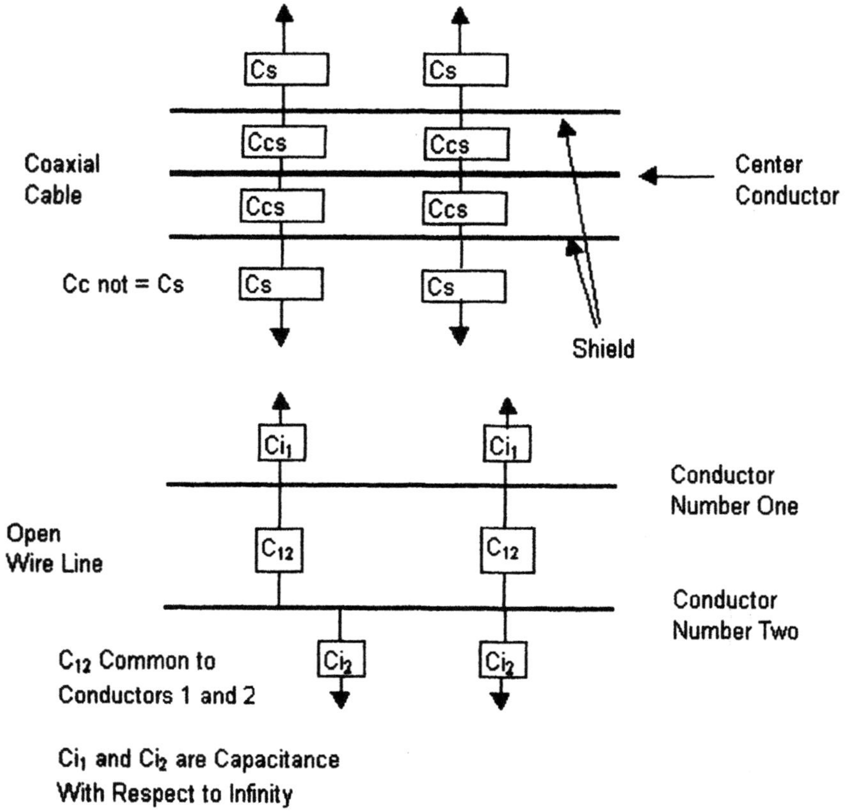

Figure 4.30. **Coupling Of Cable Conductors**

The goal of shielding on the wires is to prevent the energy from being picked up by the wiring. This energy can be primarily either E fields, or H fields.

One of the simplest shields to implement is the shielding that is provided by coaxial cable. The reason that this is easy to implement is that the shield on the coax cable provides shielding against E–fields, and the fact that the two conductors are close to each other, minimizes the effective loop area between the center conductor and the shield (thus providing H field shielding). In the case of electric field shielding, all that needs to be done is to connect one end of the shield to the return line. Only one end of the shield is needed to be connected if the length of the cable is less than 10% of the wavelength of the frequency that is causing the problem. If the shield is longer than 10%, then several of the points along the coax may be required to be connected. In the case of magnetic coupling reduction, it is always necessary to connect the return lines to both of the ends of the shields in order to provide a return current path, and thus cancellation of the magnetic field.

The way to model how this is effective for E field is by creating an equivalent model for the capacitive coupling values. In a similar method, a model for the equivalent model of the inductive elements can also show how the shielding is effective for H-fields.

4.6 SHIELDS

4.6.1 Purpose of shields

A subject that is not well known with respect to the mathematical and quantitative analyses process, yet is intuitively known is the concept of shielding for EM fields. Shields are a simple concept and in EMC serve two functions:

- Keep emissions inside
- Keep external sources/energy outside

As most people are aware, the role of a shield is to isolate fields from components that are operating. Shields can be used to provide protection against both types of EM fields, electric and magnetic. Recall that voltage causes electric fields and current causes magnetic fields. A metric used in shielding work is the measure of how well the shield does its job. This is expressed in terms of shielding effectiveness, abbreviated as SE. SE is measured in dB and is used to measure the amount of isolation to field strength levels. It is typical for good shields to provide 100 or more dB of SE. This means that the level of the field strength (for electric fields) on one side of the shield may be over five orders of magnitude greater than that on the other side! The challenge is that in order to achieve these high levels of SE, especially at very high frequencies, the shield must not have any holes, slots, or openings that will compromise its integrity. A test chamber for EMC may need to have extensive interlocking pieces to provide this level of SE, and is part of the reason why these chambers are expensive to acquire and maintain. Another complication, which is being experienced more and more frequently because of the increase in speed for digital computers, is that openings are required in the enclosure to attach peripherals, power cables, and I/O devices. Even small openings for those functions can compromise the effect of the best shielding.

4.6.2 Shielding effectiveness

Let's examine the metrics and calculations involved in the determination of shielding effectiveness. As was stated earlier, SE is expressed in dB and can be calculated as the ratio of the field strength on one side of the shield and the field strength on the other side of the shield. For example see Figure 4.31:

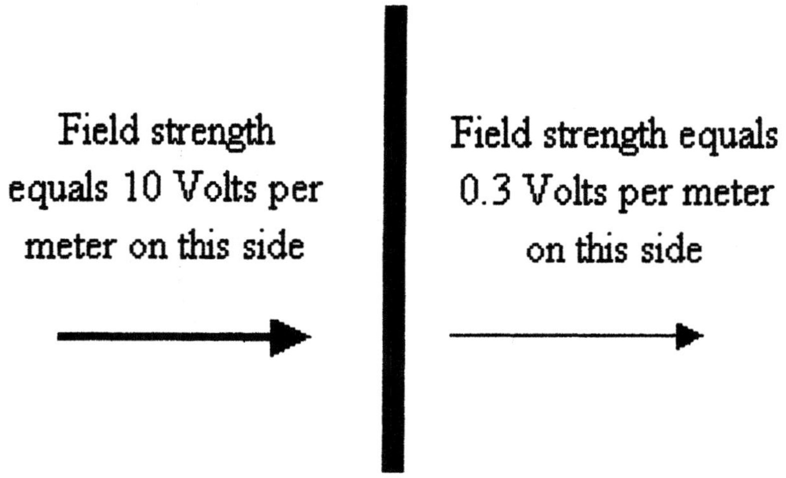

Figure 4.31 **Shielding Effectiveness**

The SE would he the following:

SE = 20 log (10 / 0.3) = 29.6@ dB

What this means is that there is a 30 dB reduction of field strength because of to the shield.

The actual process that takes place in shielding consists of two main items:

- The first is the reflection of the incident field
- The second is the absorption of the energy within the shield material.

The relative contribution of each one of the mechanisms is dependent on whether the field is electric or magnetic, and low or high frequency. This is shown in Figure 4.32.

This means that the source of the energy is on the left side of the shield, and the device to be protected is on the right side of the shield. For electric field shielding (at low frequencies), the reflection is the primary cause of the SE, and at high frequencies, absorption of the energy occurs.

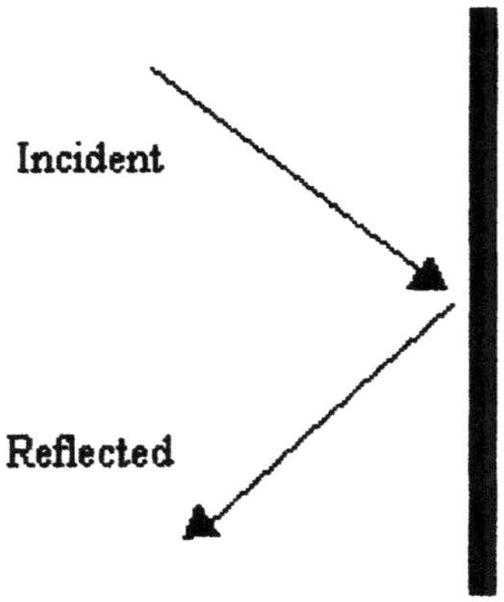

Figure 4.32 Reflected **Wave From Shield**

4.6.3 Key parameters in shield design (electric field)

Important parameters include the thickness of the material sometimes known as the barrier thickness. It is important to know what this is with respect to the skin depth at the particular frequency of concern. If the thickness of the material is equal to or much greater than the skin depth, then there is attenuation within the material. If the thickness is equal to or less than the skin depth, then the primary source of the SE is the reflection at the interface between the field and the material. This is shown in Figure 4.33:

It turns out that the thickness is important to the magnetic H field shielding capability of a material. This is because of the attenuation that takes place as the H field is passing through the material. Attenuation occurs because the magnetic field induces current in the material (a conductor), and these currents flow in a circular pattern. This pattern is similar to those seen in water, and are called Eddy currents. See Figure 4.34.

Another aspect is that these circulating currents also produce heat, due to the I^2R losses (this is an easy way to tell if a transformer is working is by feeling if it's warm!)

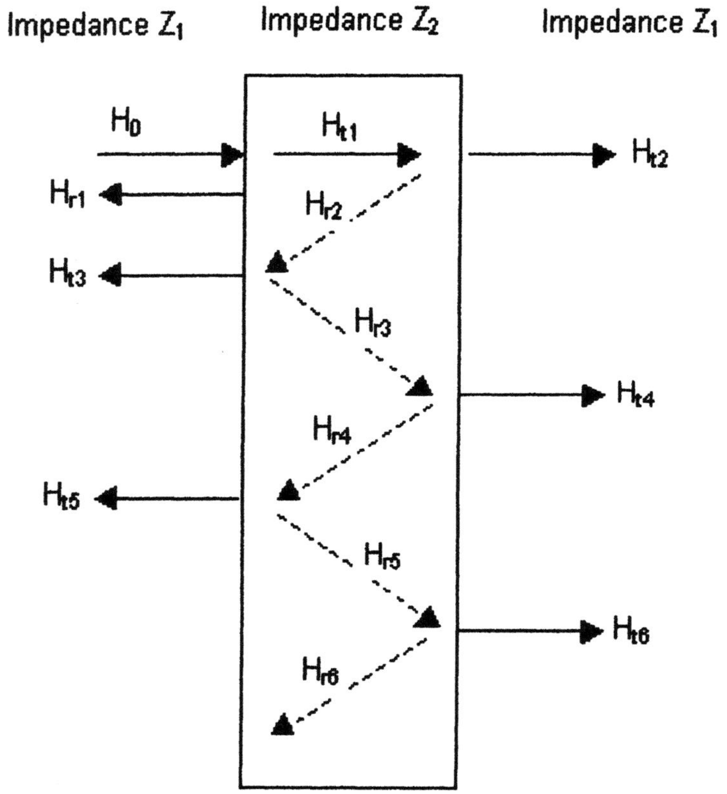

Figure 4.33. **Reflections In A Thin Shield**

The difficulty with shields is that they must be constructed and maintained to ensure their integrity. If there are openings in the shield, or discontinuities in the shielding, this can result in a path to the device or component that was intended to be protected.

We can calculate the size of openings that will allow energy to pass through. These openings are related to the wavelength of the energy.

In Figure 4.35, if D is the length of the shield, and if D is much greater than the wavelength of concern, then there will be significant shielding. If an opening is in the shield, an opening that is equal to ½ of a wavelength of a particular emission will act identical to a one-half wave antenna, as well as harmonic frequencies of the emission.

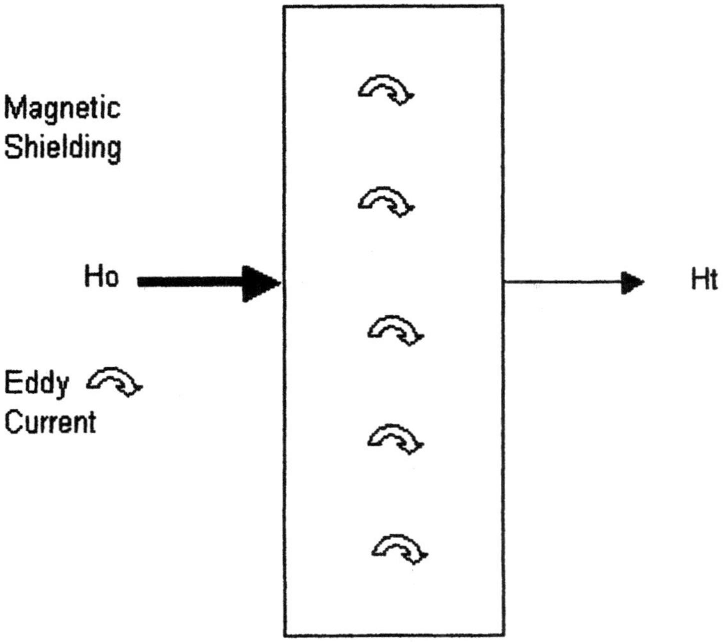

Figure 4.34. **Eddy Current Loss In Shield Material**

However, this property can be used to an advantage in applications where traditional antennas could not be used, such as on aircraft. In that case the antenna above the surface of the aircraft could be broken off, or affect the airflow over the surface. It may be possible to design the shield and then cut an opening in it, which would be covered by non-shield material. This would act as an efficient antenna. Typical usage of these types of antennas are the VHF and above frequencies.

Look at a practical example of this situation in Figure 4.36. Suppose we have a completely shielded enclosure, and place two radio receivers in the enclosure. One of the radios receives signals in the AM broadcast band (about 1 MHz) and the other one receives signals about at 500 MHz. If there are both 1 MHz and a 500 MHz signals external to the enclosure, because the receivers are in the enclosure, neither one will receive a signal.

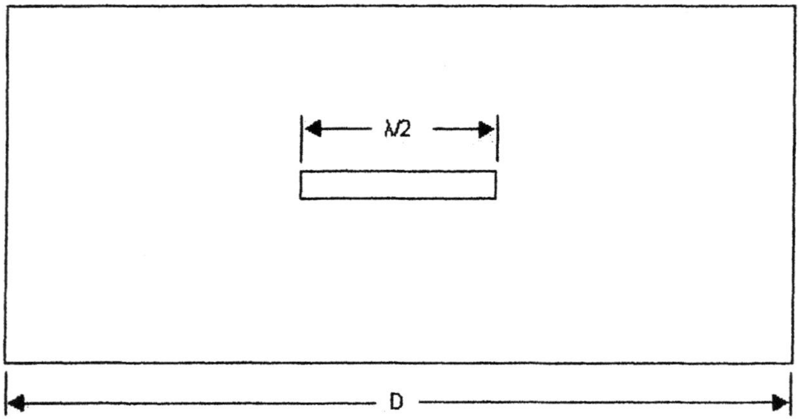

Figure 4.35. **Opening In A Shield**

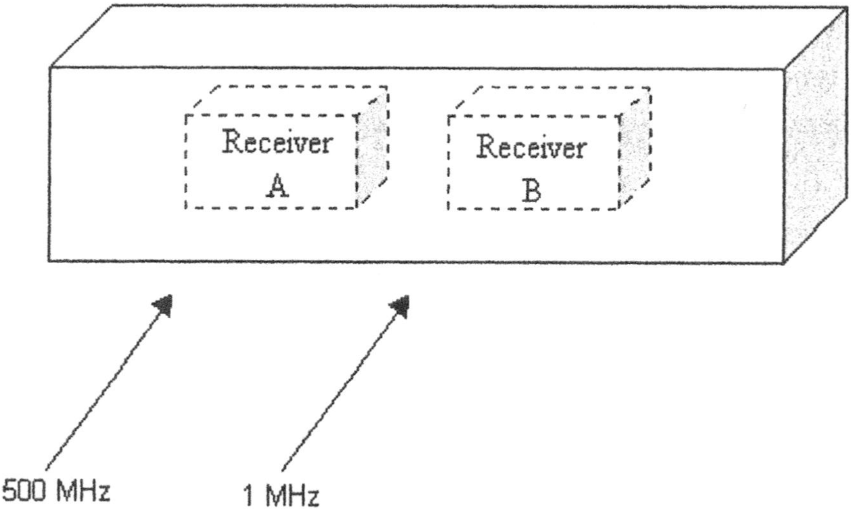

***Figure 4.36.* Receivers Inside a Shielded Enclosure**

Now if we cut openings in the enclosure of various lengths, as in Figure 4.37, one of three-quarter meter, 1 meter, and 2 meters, one of the receivers will operate. This will be the 500 MHz receiver, and the 1 MHz receiver will still be mostly unaffected by the openings. The reason is that the wavelength of the 500 MHz signal is 3/5 of a meter, which is smaller than even the smallest opening. Therefore, the signal will pass through.

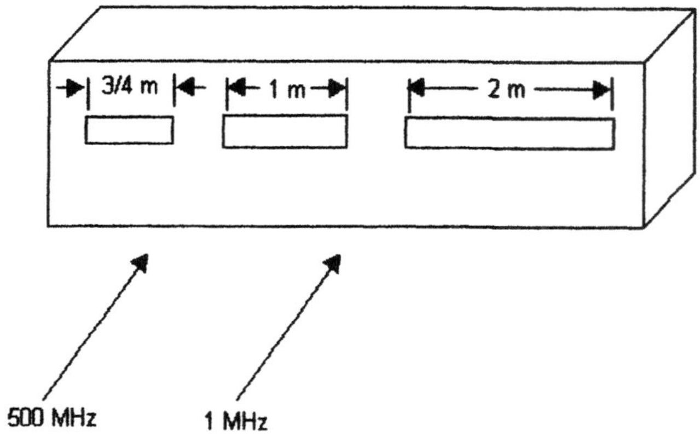

Figure 4.37. **Receivers Inside Shield With Openings**

A familiar example can be seen when travelling in an automobile, and going underneath expressway bridges or through tunnels. (Figure 4.38) AM radio stations fade out, yet the FM stations are seemingly unaffected. This is because the distance from the vehicle to the bottom of the bridge or tunnel (the opening) is much greater than the FM signal wavelength (about 3 meters). The AM radio signal is about 300 Meters, much greater than the dimension of the vehicle to the bridge or tunnel.

The visual and easiest visualization of this condition is that of a gate in a fence. (Figure 4.39). When the gate is opened, only those items that are smaller (the wavelength) than the opening can easily move through the gate.

This is why portable AM radios do not work in automobiles, yet cellular telephones, pagers, and other devices that use very high frequencies (typically UHF) work. Although this analogy illustrates a point, this is not actually what occurs.

The concept of maximum linear dimension is an important one consider when determining effectiveness of a shield. In Figure 4.40 we should understand the concepts that are relevant to this issue, that is, which opening provides the most amount of shielding. We see that we have two different types of openings, and according to our discussion of shielding effectiveness; the longest linear dimension would be the least effective shielding. This means that the rectangle would be the least effective, even though the area of the circle may be more.

Figure 4.38. **Vehicle Tunnel**

In addition to the rectangle providing an opening for the energy to be transmitted through, there's a "rule of thumb" that states:

dimension ~ Wavelength / 20

for enough energy to transfer so that a non-zero field strength will be developed.

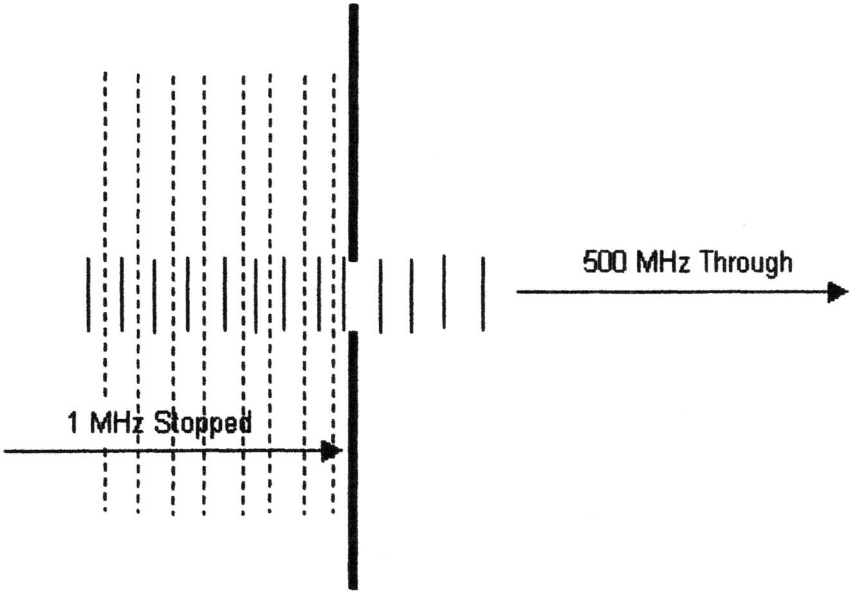

Figure 4.39. **Gate In A Fence**

For another issue related to shield effectiveness, observe the example of an assembly with a top that is fastened to it and sealed with a gasket that is intended to be compressed between the top and the rest of the assembly. An electrical model of this can be developed. It would comprehend that we have capacitance between the top, and the rest of the assembly. Because we have capacitance across this interface, we will also have some voltage differential across this interface. We can model this by adding some parasitic capacitors as in Figure 4.41. Since we have a voltage, there may be current flow and then radiation of energy can occur. The goal is to keep all the surfaces at the same potential, which eliminates any current flow, thereby eliminating radiation of energy. This can be accomplished through tight coupling between the top and the bottom. This tight coupling results in a high as possible capacitance between the two conductive surfaces. If we look at the equation for reactance of a capacitor, this makes sense because when there is an increase in capacitance the reactance decreases, which means that be voltage drop will be less for any given amount of RF current, and the emissions would be less.

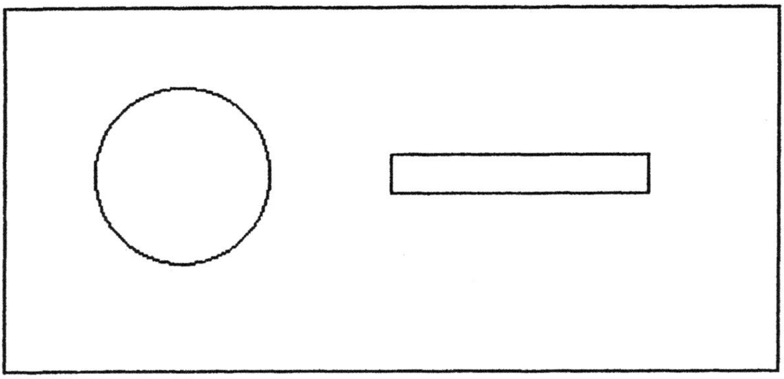

Figure 4.40. **Slot And Round Hole In Shield**

Figure 4.42 shows an important tool that can be used when working on EMC problems. This is the use of material called "shielding tape". The shielding tape would be applied along the openings in seams of assemblies to determine if the voltage differential across those interfaces is responsible for radiation. The shielding tape is available in many different widths and shielding capabilities, and may provide attenuation from a minimum of tens of dB's to close to 100 dB of isolation, and as a last resort (when commercial shielding tape is needed, and not on hand, the material may actually be fabricated from household "aluminum foil"! One of the authors of this text has personally used this method to provide suppression of radiated emissions, as well as immunity to external fields.)

Let's look at the practical aspects of providing good shielding with actual components and interfaces (such as wires or cabling). In Figure 4.43, if we have an assembly with a cable going to it, we would seek to ensure that we have an unbroken shield between the cable and the assembly itself. If we are not concerned with EMC, there is any number of different ways to run cables in and out of the assemblies; however, from an effective EMC standpoint, we want to provide an unbroken shield between the wire and the assembly itself. This could be accomplished by using coaxial cable with the outer shield of the coaxial cable being electrically and physically connected to the conductive surfaces of the assembly. This allows us to maintain a high degree of shielding and still have electrical connections needed for the signals or power to be connected to the assembly. This is shown in Figure 4.44.

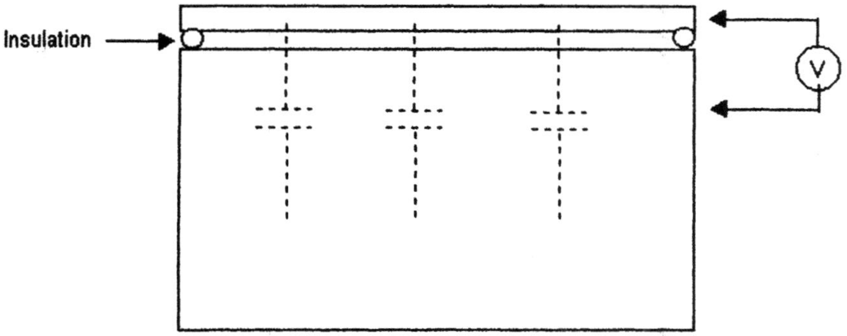

Figure 4.41 **Shield Assembly With Gasket**

In automotive EMC are there any times when application of shields are actually utilized? Although not a preferred practice because the automotive environment is never favorable to permit good application of shields in a production environment. Long-term durability and serviceability are also issues. The reason is that many times shields installed incorrectly, become damaged in use, or are not even reinstalled in the service environment. There may be times, however, when shields are required to address a radiated emissions EMC issue. One such example is the suppression of noise from the ignition system, and the way that this would be done would be to surround the secondary wire where it interfaces to the spark plug and to establish an electric field shield. This is shown in Figure 4.45. The electrical model of this is shown in Figure 4.46

What is the goal of the shield in this application? The goal is to reduce RFI from the ignition system, which possibly may interfere with the vehicle entertainment system or other devices on the vehicle. The method to reduce this radio frequency interference would be to use some type of shield around the spark plug, which is the source of the emissions. Using a highly conductive material would serve as this electric field shield.

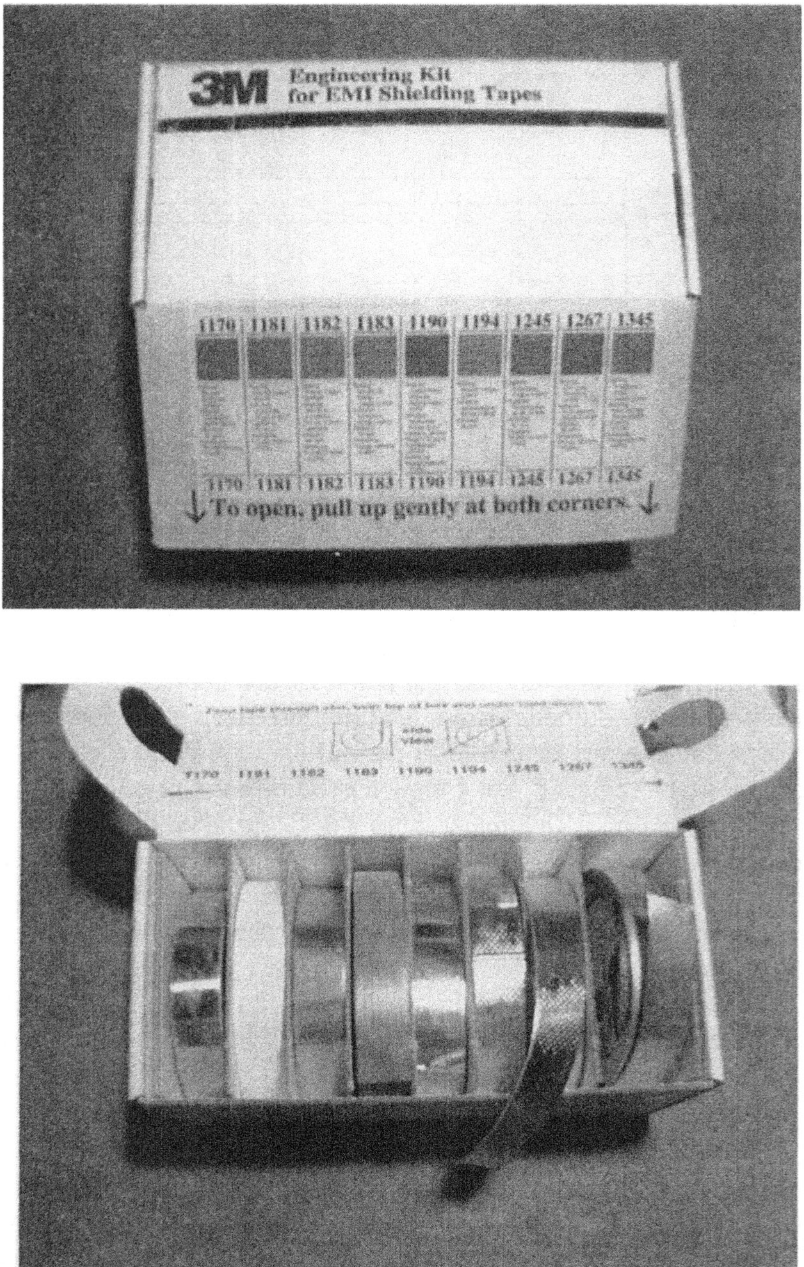

Figure 4.42 **Shielding Tape**

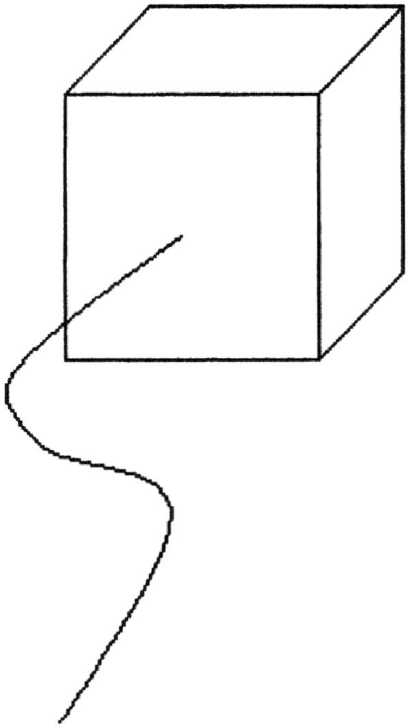

Figure 4.43. **Assembly With Cable**

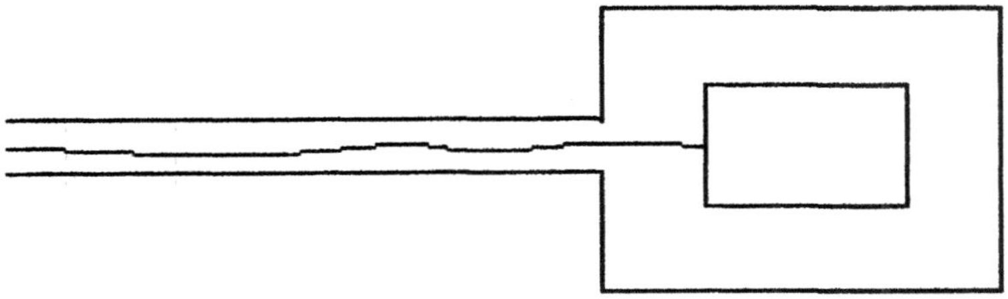

Figure 4.44. **Shielding Goal With Cables**

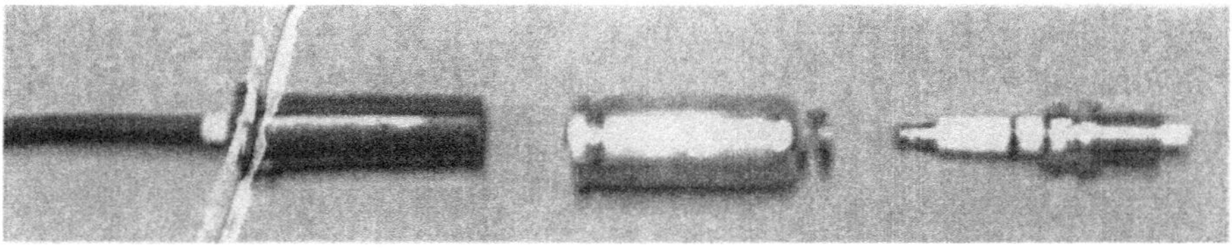

***Figure4.45.* Spark Plug Wire With Shield**

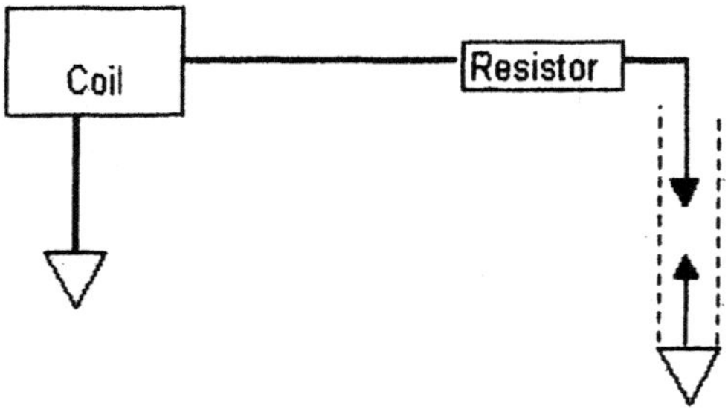

***Figure4.46.* Equivalent Electrical Model**

However, there are several key issues involved that must be remembered when designing or incorporating this type of corrective action in the automotive environment. They are as follows:

1. The shield must maintain electrical connection to the engine block (and return) or in fact the shield function may actually turn into a "parasitic" radiator. What this means is that the material itself may actually be re-radiating the energy it was intended and designed to suppress. Rather than being a solution to a problem, another problem could be created that may be worse than the original situation!

2. It is important also understand at this point that this is really not desired as an approach (since it is the "last chance"). It does depend upon integrity of the manufacturing process and long-term durability of the shield itself, and that the service procedures will install the shield later. Although clearly not preferred methodology, it may be sometimes be of use in shielding the automotive EMC environment

when required or desired.

At this point we can summarize the process of shielding for electric and magnetic fields. Refer to the following diagrams for a visual understanding of how each one of these shielding mechanisms works.

From the understanding of electric field shielding, it is important understand whether we are near field or the far field (which is discussed another chapter), which will determine the primary shielding mechanism. They are:

1. At high frequencies, the primary electric field shielding methodology is absorption.
2. At low frequencies, the primary shielding mechanism is that of reflection.

Magnetic field shielding is primarily a function of absorption of the energy within the shield. The goal of magnetic field shielding is to provide a low reluctance path for the magnetic field. There are several detailed studies of magnetic field shielding parameters; the reader is encouraged to review those materials if interested in the subject. In most automotive EMC applications, little work needs to be done to minimize magnetic field shielding. Therefore, those details of magnetic field shielding techniques will not be discussed. This discussion has been done here to acquaint the reader with the concepts. This also shown in the following diagram.

4.7 FOURIER SERIES AND FREQUENCY SPECTRUM ENVELOPE

Every periodic signal can be represented in the time domain by a technique known as the "Fourier series expansion". This explains the conditions shown in Figure 4.47:

This figure shows that a periodic signal (in this case a square wave) is a summation of sinusoidal signals of multiple frequencies and amplitudes. In a perfect square wave (50% duty cycle) only odd harmonics are present. If the duty cycle is not 50%, all harmonics will be present (shown in the last line of the figure). This is shown in the first part of Therefore, the signal has corresponding representation in the frequency domain. A given signal (e.g., a square wave with finite transition times) occupies a frequency spectrum.

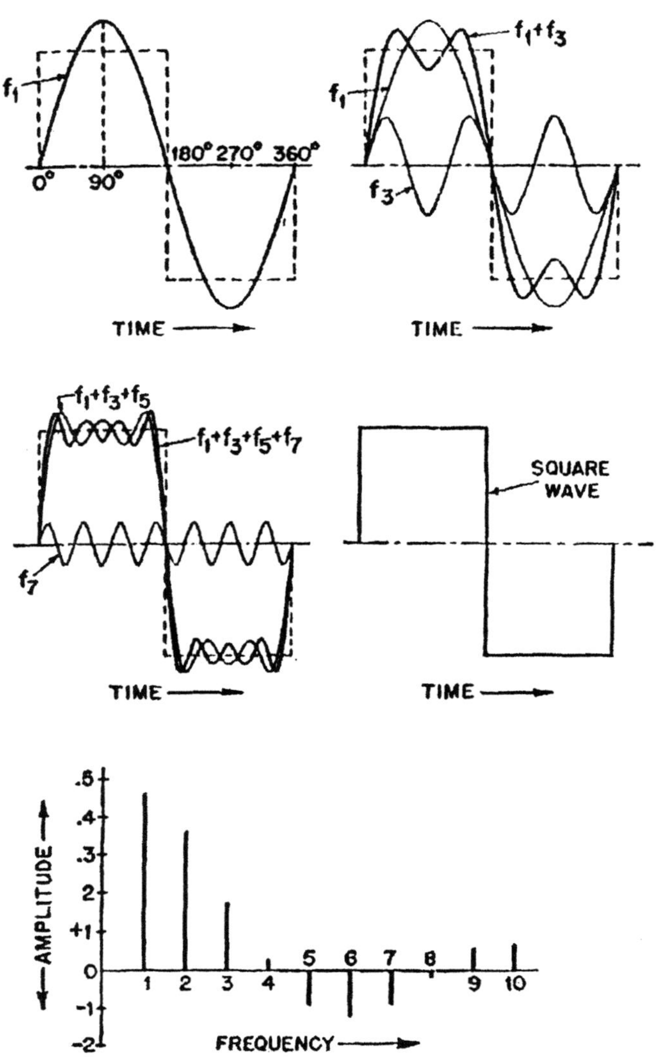

Figure 4.47 **Frequency Content of Square Wave**

In the interest of time and practicality, the Fourier envelope approximation method is used to quickly calculate the worst-case frequency spectrum envelope. For a given periodic square signal with finite rise and fall times, shown in Figure 4.48, the frequency spectrum envelope is calculated knowing:

- Peak amplitude *A* (volts, amperes)
- Pulse width t (measured at half-max)
- Period *T*
- Rise time *tr* for transition from 0.1 to 0.9 A.

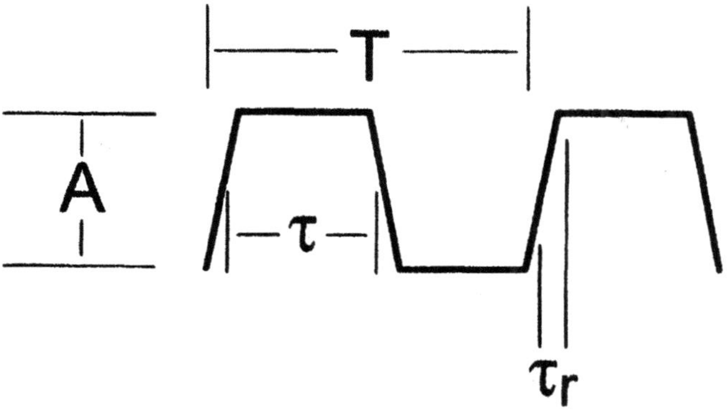

Figure 4.48. **Periodic Square Wave Signal**

The frequency spectrum envelope shown in Figure 4.49 is calculated using the above information and equations derived from the trigonometric Fourier transform. Amplitude of the signal in frequency domain (Af) is calculated using:

$$Af = 2\,A\,\tau/T\,, \quad (3\text{-}5)$$

where *A* is peak amplitude in the time domain. Corner frequencies, f1 and f2, are calculated using the equations:

$$f_1 = 1/\pi\tau$$

and

$$f_2 = 1/\pi t_r\,.$$

It should be noted here that in practice the signal waveforms are not completely symmetrical. In this case, it is important to use the faster of the two transition times, the rise time or the fall time. Figure 4.48 shows that between the first corner frequency, f_1, and the second corner frequency, f_2, the amplitude decreases at a rate of 20 dB per decade of frequency.

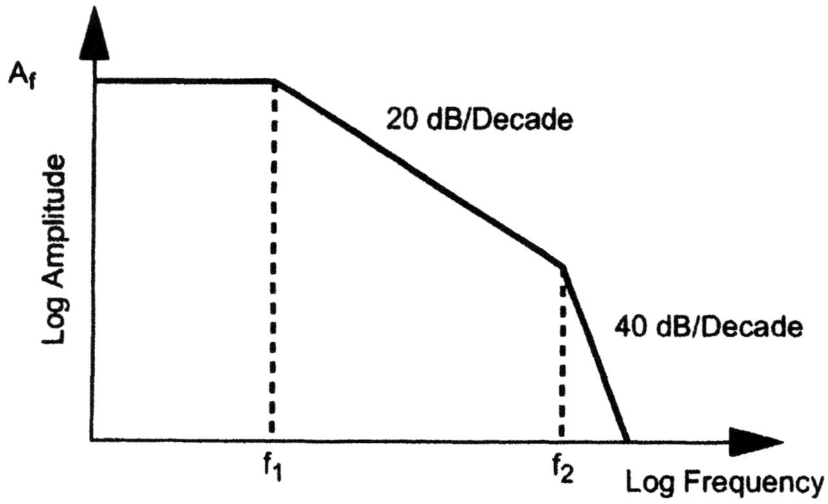

Figure 4.49. **Frequency Spectrum Envelope**

At frequencies above f_2 , the amplitude decreases at a rate of 40 dB per decade of frequency. Figure 4.50 shows a frequency spectrum envelope overlaid upon an actual frequency spectrum.

While this method does not yield an exact frequency spectrum plot, the resulting frequency spectrum envelope does provide a worst-case envelope for a given time-domain signal and other important information. Changes in duty cycle and transition times reduce the frequency spectrum envelope. For a 5-Vp , 500-kHz signal with a 50-percent duty cycle and transition times of 10 ns, f_1 is 318.3 kHz, f2 is 31.8 MHz, and Af is 5 Vp . By changing the duty cycle to 30 percent and the transition times to 100 ns, f_2 becomes 3.18 MHz and Af becomes 3 Vp . This implies that noise amplitudes are reduced and noise frequency amplitudes lowered (Table 4.6).

4.8 CAPACITORS, INDUCTORS, AND ACTUAL PROPERTIES

In designing the filter, it is important to note that the capacitor or inductor being used is not an ideal component and will not act as such. A capacitor, even the leadless surface mount type, exhibits parasitic inductance and resistance. "Parasitic" describes the capacitances and inductances that do not appear on engineering drawings, but nevertheless exist and cause odd things to happen to the desired signal or waveform. The term "stray capacitance" is a commonly used term that means the capacitance between a conductor and its surroundings. A good example of "stray capacitance" is between a switching transistor and the heat sink upon which it rests, typically 50 to 150 pf. As a general rule, when trying to bypass a certain frequency, try to keep the reactance of the capacitor being used around 0.1 Ω. A lower reactance (0.01 Ω) may tend to self-resonate.

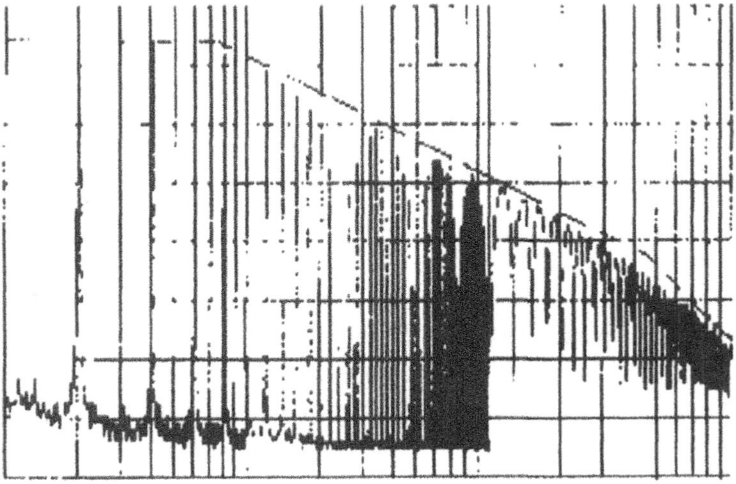

Figure 4.50. **Frequency Spectrum and Frequency Spectrum Envelope**

Table 4.6 **Frequency Spectrum Envelope Calculations**

A	**T(1/f)**	**τ (T X Duty Cycle)**	**τ_r**	**Af**	**f_1**	**f_2**
5V	2 μs (1/500 kHz)	1 μs (2 μs X 50%)	10 ns	5V	318 kHz	31.8 MHz
5V	2 μs (1/500 kHz)	0.6 μs (2 μs X 30%)	100 ns	3V	531 kHz	3.18 MHz

Figure 4.51 shows models of a capacitor and an inductor and includes parasitics. The capacitor parasitics are lead and plate resistance and inductance, dielectric losses, and skin effects losses.

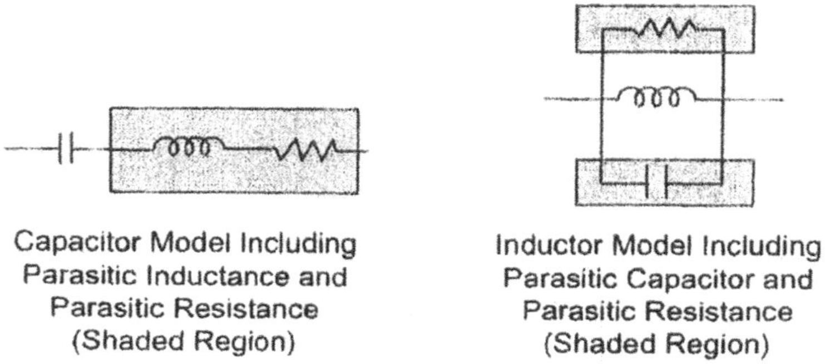

Figure 4.51. **Capacitor And Inductor Models Including Parasitics**

The inductor parasitics are lead and winding resistance, turn-to-turn and turn-to-core capacitance, dielectric losses of insulation, eddy current losses, hysteresis losses, and skin effects losses. One consequence of parasitics is that they cause the inductor or capacitor of a filter to self-resonate at its resonant frequency (100 kHz to 20 MHz for capacitors and 2 to 100 MHz for inductors) and create EMI problems. Another consequence is that the impedance of the inductor or capacitor is non-ideal above the frequency which the parasitic components begin to have an appreciable impedance (Figure 4.52.)

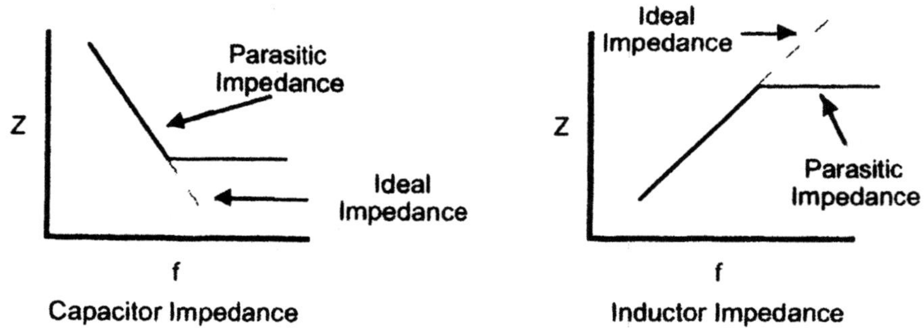

Figure 4.52. **Inductor and Capacitor Impedance**

4.9 FILTERING OVERVIEW

As stated before, the purpose of the EMI filter is to prevent the entry or exit of undesired electromagnetic energy from equipment. A filter absorbs the noise energy through the use of lossy elements such as resistors and ferrite components, or reflects the noise energy back to the source through use of reactive elements. Generally, EMI filters are low-pass filters with effectiveness depending on the impedances of the elements at either end of the filter.

For a filter that attenuates EMI by reflecting noise, the filter should provide a maximum impedance mismatch. If the load impedance is low, the impedance of the filter from the load viewpoint should be high. If the load impedance is high, the impedance of the filter from the load viewpoint should be low. Figure 4.53 gives filter configuration examples for various load and source impedances.

EMI filters are single-section filters or several single-section filters cascaded together for more attenuation. It has been demonstrated that a two-section filter has a lower optimum weight than a single-section filter when by design both have identical filtering properties. The number of sections and configuration are not limited to this presentation. It is important to remember to isolate the input and output cables of the filter. Isolating input and output cables from each other prevents the cables from coupling to each other and bypassing the filter. Isolation may be accomplished by placing the input cables and the output cables on opposite sides of the filter. However, to properly isolate the cables and prevent noise from bypassing the filter, the filter may have to be shielded by placing it in a shielded enclosure.

4.9.1 Common Mode Filtering

The various filter configurations shown in Figure 4.53 are DM filters. The other type of conducted noise, CM noise, requires a different type filter. CM filters are usually CM chokes or line-to-ground filters such as feed-through capacitors. The CM choke relies on the magnetic properties of ferrite cores to absorb CM noise.

Figure 4.54 shows a schematic of a multi turn CM choke. The cables are wrapped four to five turns around a ferrite core. The magnetic field (Hdm) induced by the DM current (Idm) on one side of the core is canceled by the magnetic field induced by the DM current on the return side of the core. Therefore, the DM current is not attenuated. However, for the CM current

(Icm) the magnetic fields (Hcm) do not cancel, and the series combination of the inductive reactance and resistive losses of the core attenuate the CM noise. Figure 4.55 shows CM choke configurations and shielding.

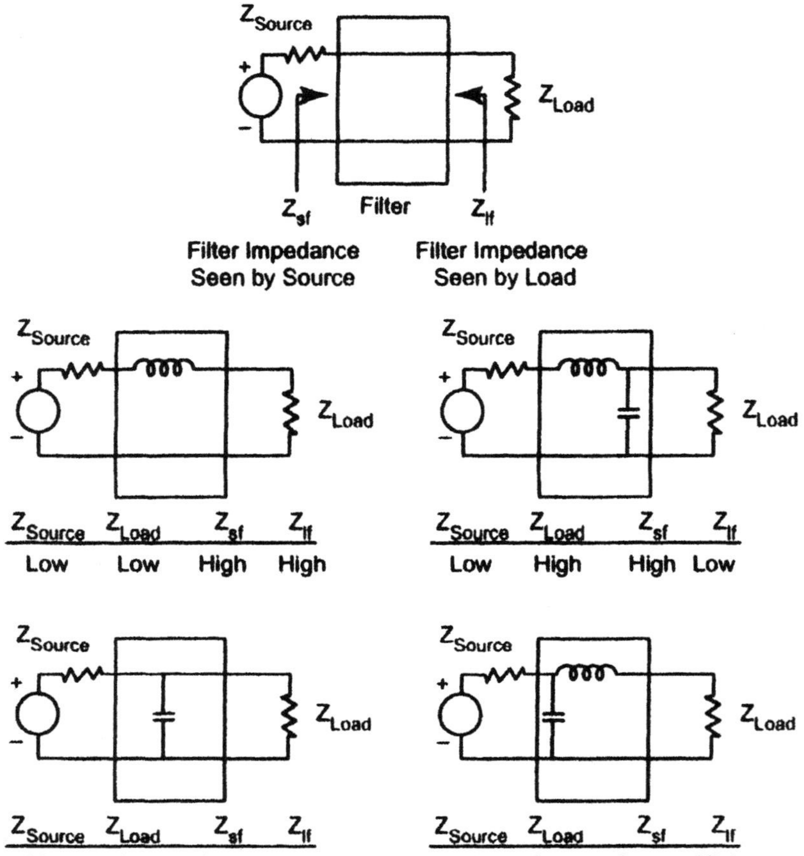

Figure 4.53. **Filter Configuration Examples**

4.9.2 Isolation

Isolation is another means of diverting undesired electromagnetic energy. Two methods commonly employed are isolation transformers and opto-isolators. The isolation transformer may be used in ac power circuits, in switched-mode power supplies, and in analog signal circuits such as data lines. The isolation transformer breaks up the return loop by increasing its impedance. Figure 4.56(a) shows the schematic of a typical isolation transformer. At low frequencies, the capacitance between the primary and secondary windings presents a high impedance in the conducted path. At high frequencies, however, this capacitance impedance is no longer substantial and does not appreciably attenuate CM or DM noise. Addition of a Faraday shield between the primary and secondary windings attenuates high-frequency noise. The primary-to-secondary capacitance is divided between the primary winding and shield and between the shield and

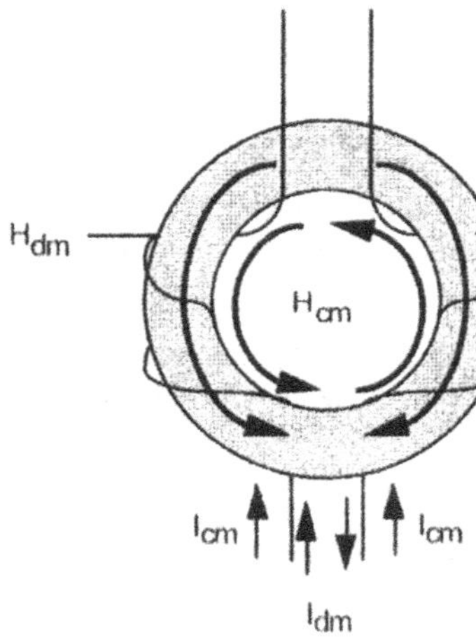

Figure 4.54. **Multi-Turn CM Choke**

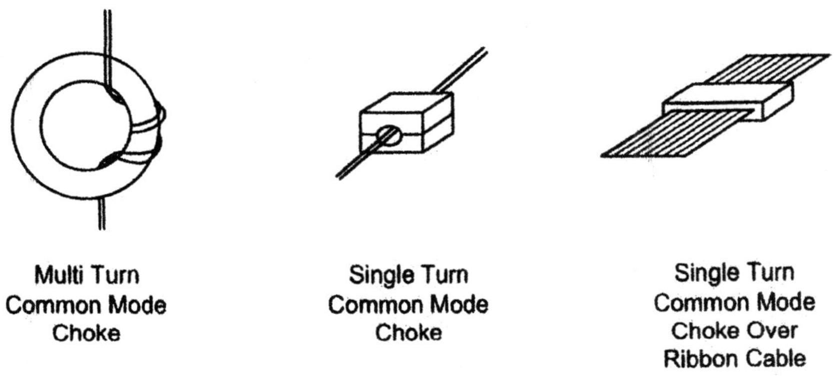

Figure 4.55. **CM Choke Configurations**

secondary winding. For CM reduction, the shield is connected to the transformer housing that is connected to ground. This ground connection impedance, along with the winding to shield capacitance, acts as a voltage divider to reduce CM noise coupled across the transformer. For DM reduction, the shield is connected to the return side of the transformer to short-circuit the DM currents. Figure 4.56(b) shows the schematic of a Faraday shielded isolation transformer for CM reduction. Figure 4.56(c) shows the schematic of a Faraday shielded isolation transformer for DM reduction. Figure 4.56(d) shows the schematic of a triple Faraday shielded isolation transformer that provides common and DM isolation from either side of the transformer.

Opto-isolators are another method of isolating signals to attenuate conducted EMI. Figure 4.57 shows a schematic of an opto-isolator. Opto-isolators perform over a wide bandwidth (approximately 50 MHz) and work with both logic and analog signals above 100 mV. The limiting factor in high-frequency usefulness of opto-isolators is their input-to-output capacitance (typically 0.1 to 10 pF). This capacitance allows high-frequency noise to bypass the high impedance of the opto-isolator.

4.10 ENCLOSURE SHIELDING

Most books on shielding delve into a comprehensive coverage of shielding theory that is beyond the scope of this handbook. Guidelines provided here provide a minimum of mathematics and theory. Shielding of EM fields is accomplished through reflectance or absorption of the fields by a barrier. In most applications, the barrier is a metal, although coated and conductive plastics are being used more frequently in commercial applications. An important point to remember in shielding is that the actual shielding provided by a metal barrier depends on the type electromagnetic field that predominates. Reflection is highly effective against predominately electric fields and plane waves and has little effect on predominately magnetic fields.

Absorption is the mechanism in predominately magnetic field attenuation. Reflectance increases with surface conductivity of the shield and decreases with frequency. Absorption increases with:

- Thickness of the shield
- Conductivity of the shield
- Permeability of the shield
- Frequency of the incident field

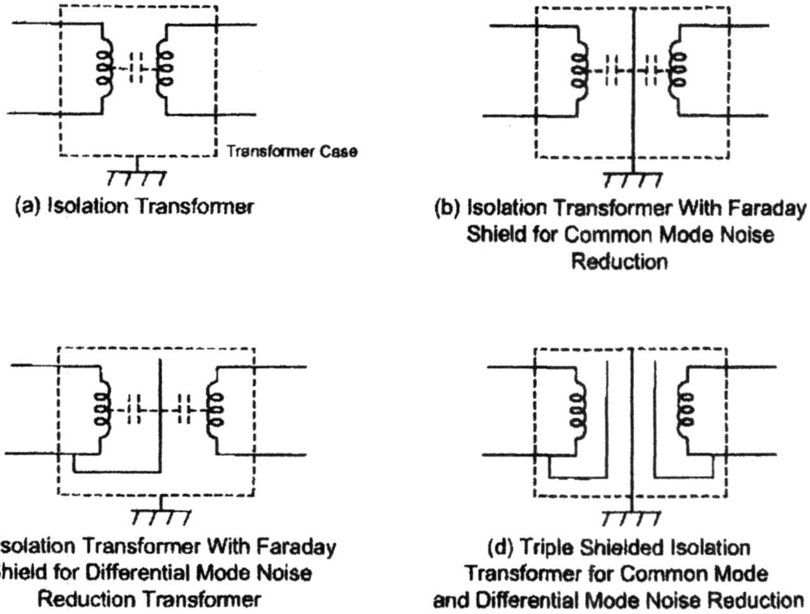

Figure 4.56. **Isolation Transformer Configurations**

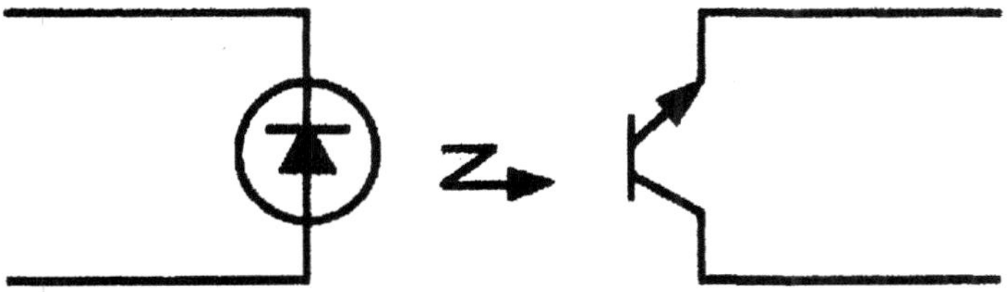

Figure 4.57. **Opto Isolator Schematic**

Absorption in a metal barrier is exponential in nature, i.e., as an electromagnetic field passes through a metal barrier, the amplitude of the electromagnetic field decays exponentially. At some distance into the metal barrier, the amplitude of the impinging electromagnetic field has decreased to 1/e or 33 percent of the amplitude at the surface of the barrier. The distance at which this occurs is called the skin depth of the metal. The formula for skin depth is given in equation

$$d = 2.6 / \sqrt{\mu r \delta r f MHz} \quad \text{in mils}$$

where μr is the permeability of the metal relative to copper,
δr is the conductivity of the metal relative to copper, and
f_{MHz} is the frequency of the electromagnetic field impinging on the metal.

The skin depth concept is shown in Figure 4.58. Table 4.7 lists plane-wave skin depths for copper and aluminum at various frequencies.

The performance of a shield in reducing the electromagnetic energy that passes through it is known as its shielding effectiveness. The equations below defines the shielding effectiveness (in decibels) for electric fields and magnetic fields:

$SEdB = 20 \log10 \{Ein/Eout\}$ for electric fields
$SEdB = 20 \log10 \{Hin/Hout\}$ for magnetic fields

where Ein (Hin) is the field strength incident on the shield, and Eout (Hout) is the field strength after passing through the shield.

Shielding effectiveness is shown in Figure 4.59. Note: At one skin depth the *SE*dB of a metal is at least 8.7 dB and at 2.3 skin depths the *SE*dB is at least 20 dB.

The above discussion assumes that the barrier or shielding material is homogeneous and large such that there is no leakage or edge effects. The shielding effectiveness expressed in the equation is degraded by apertures for connectors, switches, and I/O lines and seams for doors, access panels, and cover plates. These apertures and seams serve as leakage paths for electromagnetic energy; this leakage lowers the SE_{dB} of the barrier.

Finally, a few shielding rules of thumb:

- For a predominately electric field or plane wave, use a good conductor (copper or aluminum) to maximize reflection loss.
- For a high frequency magnetic field (frequency >500 kHz), use either a good conductor or a material with a high permeability, μr.
- For a low frequency magnetic field (10 kHz > frequency > 500 kHz), use a magnetic material such as steel, for frequency <10 kHz use a material with a high permeability, μr, to maximize absorption loss.
- Reflection loss varies with the type of field; absorption loss is independent of the field.

- A metallic shielding material thick enough to support itself usually provides good electric field shielding at all frequencies.

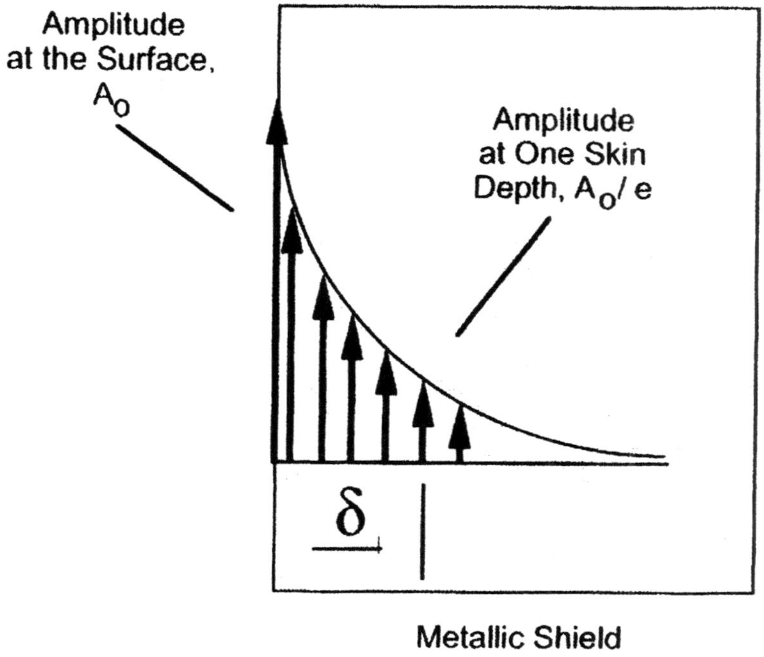

Figure 4.58. **Schematic Of Skin Depth**

Table 4.7. **Skin Depths At Various Frequencies.**

Frequency	d for Copper (mils)	d for Aluminum (mils)
10 kHz	26	33
100 kHz	8	11
1 MHz	2.6	3
10 MHz	0.8	1
100 MHz	0.26	0.3

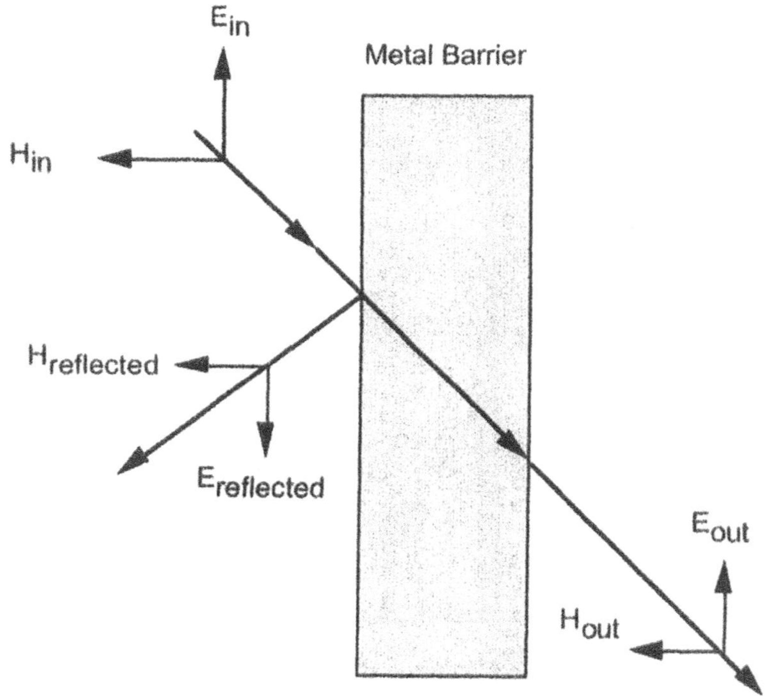

Figure 4.59. **Schematic Of Shielding Effectiveness**

4.11 Shield Discontinuities

As stated in the previous section, the shielding effectiveness of an enclosure is degraded by the introduction of discontinuities into the enclosure. These shield discontinuities are the holes, seams, and joints found in nearly all electrical and electronic equipment. Leakage through seams, holes, and joints is usually a greater concern than the shielding effectiveness of the shield material. The methods presented here are equally adequate for minimizing magnetic and electric field leakage; only the types of shielding materials differ. Discontinuity rules of thumb include:

• The amount of leakage from a discontinuity depends on the maximum linear dimension of the opening and the frequency of the source.

• A slot or rectangular hole may act as a slot antenna when the maximum linear dimension of the slot becomes greater than 1/10 of a wavelength.

• A large number of holes allows less leakage than one large hole of the same total area.

• A hole shaped to form a waveguide (the depth of the hole is greater than the diameter of the hole) can offer greater attenuation than a "regular" hole pattern for frequencies lower than the waveguide's critical frequency. This critical frequency is roughly the frequency at which the maximum linear dimension of the opening of the waveguide equals $\lambda/2$. Below this critical frequency, the waveguide attenuation is dependent on the length of the waveguide and is called waveguide below cutoff.

• For seams and joints it is necessary to maintain a continuous metal-to-metal contact along the seam or joint to ensure shielding integrity.

• The preferred seam for preventing EMI leakage is a continuous weld.

• When bolts or rivets are used to make a bond, the shielding effectiveness depends on the number of rivets or screws per linear inch, the mating pressure at the contact surface, and the cleanliness of the two mating surfaces.

• The higher the number of rivets or screws per linear inch, the greater the shielding effectiveness.

• For equipment enclosures that require ventilation, the following materials (in descending order of attenuation) should be used to cover the opening:

1. Waveguide below cutoff panels (honey-comb panels)
2. Perforated metal sheet
3. Woven or knitted metal mesh.

Chapter 5

Electromagnetic Fields

5.1 INTRODUCTION

Electromagnetic energy is transferred through propagation of waves. These waves can be described by their frequency, direction of propagation, and amplitude. When working with electromagnetic energy and physical devices, one of the fundamental concepts is that of "wavelength". Wavelength is often expressed in metric units. The definition of wavelength is the speed of the propagation of energy divided by the frequency of that energy, expressed in Equation 5.1. Wavelength is determined by dividing the propagation velocity (in meters per second), by the frequency. Radio frequency energy propagates at about the speed of light, 186,000 miles per second or 300 million meters per second. The unit of frequency is Hertz (Hz.), formerly called "cycles per second" (cps). For example, if we want to determine the wavelength of a 100-MHz signal, the wavelength would be calculated from Equation 5.1 as 300/100 MHz or 3 meters.

$\lambda = 300/f$
Where λ = Wavelength in meters
f = Frequency in MHz

Equation **5.1**

At 300 MHz the wavelength would be equal to one meter, because 300 million meters per second divided by 300 million cycles per second equals a wavelength of 1 meter. Likewise, in the AM band, a 1-MHz signal has a wavelength of approximately 300 meters and a cellular telephone signal at approximately 900 MHz would have a wavelength of approximately one-third of the meter. See Table 5.1.

Table 5.1 **Commonly Used Frequency Bands and Wavelengths**

Frequency Band	Frequency in MHz	Wavelength in meters
AM Broadcast	1.0	300
Amateur	50	6
FM Broadcast	100	3
Amateur	150	2
Amateur	450	0.66
Cellular Telephone	900	0.33

5.2 CHARACTERISTICS OF THE ELECTROMAGNETIC ENVIRONMENT

The EM (electromagnetic) environment is also defined by measuring the electrical or magnetic field component. A "field strength" will develop from a transmitter of EM energy. The amplitude of this field strength is an inverse function of distance, in free space. We measure this E (electric) field strength in terms of volts per meter or its equivalent in decibels (dB) relative to one volt per meter. For a description of dB calculations, see the appendix.

The inverse distance relationship is shown in the simple demonstration of a transmitter and a field strength meter in Figure 5.1.a. As a transmitter is feeding power to an antenna, the transmitted power will produce an electromagnetic field in space. A common way to measure this field is by measuring it with a field-strength meter. The field-strength meter will measure this field and display the field strength. As the field-strength meter is moved further away from the antenna, the field strength-indication will decrease. If the field strength meter is moved closer to the antenna, the reading will increase. A good analogy is to think about the relative brightness on a surface from a flashlight, the further away from the flashlight, the illumination per unit area is less, although the total light from the bulb has not changed.

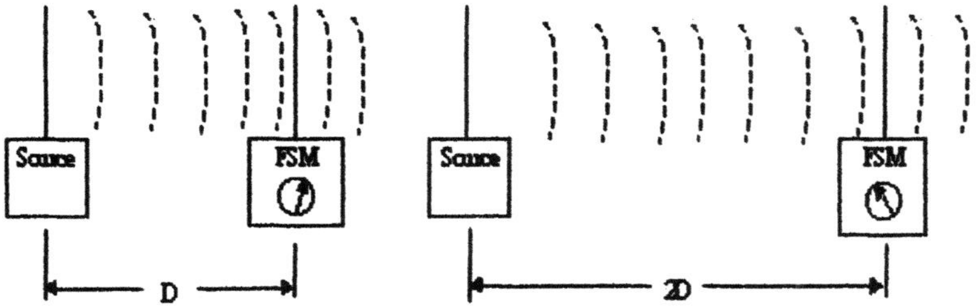

Figure 5.1a. **In the Far Field, Doubling the Distance Reduces the Field Strength to One-Half**

Figure 5.1b. **Doubling The Distance Reduces The Received Power To One Quarter**

In EMC work, electric and magnetic fields are quantified in terms of their field strength. There is significant range in the magnitude of these fields as distance is traversed. To permit readable plots of field strength versus distance, field strength is typically expressed in "decibels" (dB). The unit dB, originally used for audio purposes, is defined as ten times the logarithm of ratio of measured power to some reference power. The formula for dB with regard to reference power (P_0) is shown in Equation 5.2. The equation shows that doubling the power is equivalent to approximately a 3-dB change in level. We can easily convert this to represent voltage or current ratio relative to a reference voltage (V_0) or reference current (I_0), assuming that the resistance R is the same in all cases.

$$\text{Power (dB)} = 10 \log (P_1/P_0)$$

$$= 20 \log (V_1/V_0)$$
$$= 20 \log (I_1/I_0)$$

For the same resistance R

Equation **5.2**

EMC work uses some common units to express the voltage, current, power levels, and signals. The typical voltage units are field strength in volts per meter (V/m) or microvolts per meter (μV/m). Typical power units are watts (W) or milliwatts (mW), and current levels are amps (A), or milliamps (mA). Table 5.2 summarizes voltage, current, and power levels frequently used in EMC:

Examples of Signal Strength or Field Strength Expressed in dB:

- An immunity test field strength of 100 Volts per meter could also be expressed as 20 log (100 Volts per meter/1.0 Volts per meter) = 40 dBV/m
- A received field strength of a broadcast station of 50 μV/m could also be expressed as 20 log (50 μV/m /1.0 μV/m) or 44 dBμV/m, sometimes expressed as 34 dBu.
- A power amplifier output of 1000 watts could be expressed in dB as 10 log (1000 Watts/1.0 Watt) =30 dBWatts
- A signal generator output of 0.5 milliwatt expressed in dB could be 10 log (0.5 miliwatt/1.0 milliwatt) or –3 dBm
- And a generated test current of 10 mA could also be expressed as 20 log (10 mA/1.0 mA)= 20 dBmA

For a 50-Ohm system,

1 milliwatt = 0 dBm = 107 dBμV
-107 dBm = 0 dBμV = 1 μV

Table 5.2 **Commonly Used Voltage, Current, and Power Levels**

Amps	dBA	milliamps	dBmA	microamps	dBμA
Volts	dBV	Millivolts	dBmV	Microvolts	dBμV
0.000001	-120	0.001	-60	1	0
0.00001	-100	0.01	-40	10	20
0.0001	-80	0.1	-20	10^2	40
0.001	-60	1	0	10^3	60
0.01	-40	10	20	10^4	80
0.1	-20	10^2	40	10^5	100
1	0	10^3	60	10^6	120
10	20	10^4	80	10^7	140
100	40	10^5	100	10^8	160

dB	**Power Ratio**	**Voltage or Current Ratio**
0	1.00	1.00
0.5	1.12	1.06
1.0	1.26	1.12
1.5	1.41	1.19
2.0	1.58	1.26
3.0	2.00	1.41
4.0	2.51	1.58
5.0	3.16	1.78
6.0	3.98	2.00
7.0	5.01	2.24
8.0	6.31	2.51
9.0	7.94	2.82
10.0	10.00	3.16
20.0	100.0	10.0

The use of dB allows us the ability to express negative gain (loss) or "attenuation" such as "3 dB power loss" is equal to "-3 dB", and is calculated as follows:

Assuming that the original power is 10 watts, a 3-dB loss (negative logarithm means the larger quantity is in the denominator, the smaller in the numerator) calculation is shown in Equation 5.3.

-3 (Power Ratio (in dB)) = 10 log (X / 10)
X = 5 Watts

Equation 5.3 **Calculation of Loss in dB**

It is helpful to remember that:
-3 dB is half the power or 0.707 times the voltage,
+3dB is twice the power or 1.414 times the voltage,
-6 dB is one-fourth the power or one-half the voltage,
and +6dB is four times the power and twice the voltage.

From a practical standpoint, most EMC measurements are repeatable to plus or minus three or four dB (at best), even in the far field. As illustrated in Figure 5.2, if we have a transmitter and antenna generating a field, and measure field strength at a given distance with a field probe, we expect that the results will vary between plus or minus three or four dB when the test is repeated. The implication is that, if we measure the field strength, and calculate that the transmitter power is 10 watts, 3 dB repeatability means that the actual power may be anywhere from 5 to 20 watts. This is because of test setup variability, non-uniformity of the field, repeatability of the test instrumentation, variation in field probe and cable positioning, and in some cases errors caused by measuring in the near field of the source.

An important feature of utilizing dB is that they are additive. For example if an amplifier has 10 dB gain, and another amplifier has 13 dB of gain, and are in series with one another, we can add the gains together typical of the full system gain, which in this case is 23 dB. This is shown in Figure 5.3:

Sometimes in EMC work, "gain" antennas are employed (the antenna exhibits gain over a "reference" antenna). This means that the field strength is the same as if the transmitter were generating more power that it actually is. Antenna gain can be used rather than more powerful amplifiers to generate higher field strength. These gain antennas can also be used to measure signals or device emissions. The gain antenna will increase the signal and enable one to read signal levels that are near the measurement instrument noise floor. One disadvantage of gain antennas is that the near field region extends further outward from the antenna (because $2D^2/\lambda$ is larger), and may result in measurements being taken in the near field of the antenna, where the wave impedance is not well behaved.

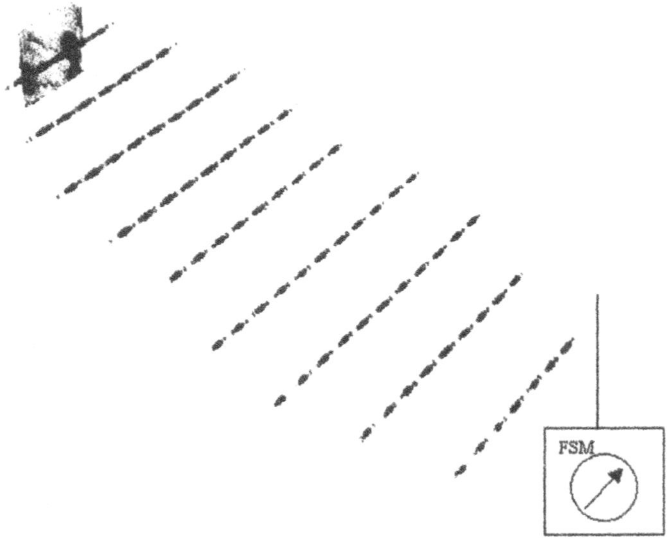

Figure 5.2. **Measuring Field Strength Generated by a Transmitter and Antenna**

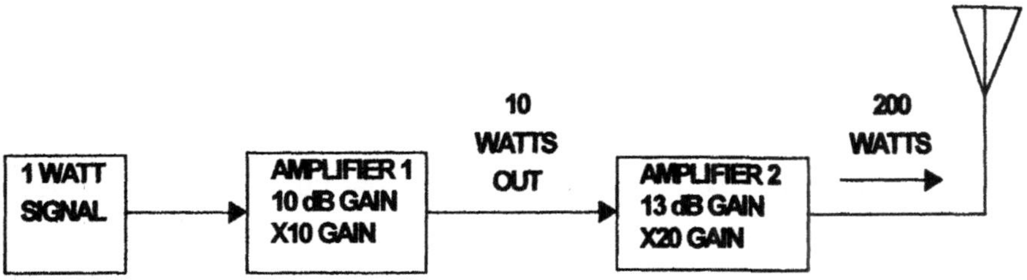

10 dB + 13 dB = 23 dB

Figure 5.3 **Addition of Decibels**

5.3 COMPARISON OF CIRCUIT THEORY WITH EM FIELD THEORY

The similarities between Ohm's law for electric circuits and free space conditions for EM fields are obvious when compared side by side as shown in Table 5.3. The comparison may help you better understand the relationship between E fields and H fields, if you already are familiar with electric circuit theory.

Table 5.3 **Comparison of Circuit Theory with EM Field Theory**

Circuits	EM Fields
Voltage potential (E), Volts	Electric field strength (E), Volts/meter
Electric current (I), Amperes	Magnetic field strength (H), Amperes/meter
Circuit Impedance (Z), Ohms	Characteristic Impedance (Z), Ohms
Circuit Resistance (R), Ohms	Impedance of free space (Z_o=377), Ohms
E (Volts) = I x Z	E = H x Z Volts/meter
When Z = resistive:	When in the far-field, Z_o = 377 Ohms:
E (Volts) = I x R	E (Volts/meter) = H x 377
and P (Watts) = E x I	and P_d (Watts/meter^2) = E x H
Substituting for E and I:	Substituting for E and H:
P (Watts) = I^2 x R	P_d (Watts/meter^2) = H^2 x 377
P (Watts) = E^2/R	P_d (Watts/meter^2) = E^2 /377

The simple free field relationships stated above apply at distances of about two or more wavelengths from the radiating source, called the far-field. Here Z (the ratio of E to H) is a fixed constant equal to 377 Ohms, and here we can determine the power density by measuring only the E field (or H field) and then calculate the power density from that number. Field survey meters (for measuring the electromagnetic environment) usually read out in terms of E^2 or H^2. Power density is E^2 (in Volts squared per meter squared) divided by 377 ohms or H^2 (in Amps squared per meter squared) times 377 ohms under free space conditions.

Note: The above does not apply when dealing in the near-field, because the Z is not usually equal to 377 Ohms. In fact, in the near-field, Z can have any value from near zero to nearly infinity, and can change very quickly from one measurement position to another. That is why both E and H must be measured when taking measurements in the near-field.

5.4 MAXWELL'S EQUATIONS

In order to understand EMC, we need to review the basic physics behind EMC. There are four basic relationships that are important to understanding the study of EMC. These are referred to as "Maxwell's equations". Although seemingly confusing and difficult to use, Maxwell's equations basically state the mathematical relationship between electric and magnetic fields and waves.

There are four basic equations that are utilized. The first one, based on Gauss's work, has to do with the concept that an electric field results from enclosing a charge. This means that if there is a charge in free space and we draw a sphere around this charge, there will be an electric field potential at some distance from this charge, and the electric field will be proportional to the total charge enclosed. This relationship can be seen in the first equation, Equation 5.4 See also Figure 5.4.

$$\partial \mathbf{E.dA} = q/\varepsilon_0$$

Equation 5.4 **Maxwell's First Equation**

The second equation, also based on Gauss's work, says that magnetic dipoles exist. What this means is that if we have a "North Pole," we must also have a "South Pole," forming a magnetic "dipole." For any closed surface the magnetic flux directed inward toward the South Pole will equal the flux outward from the North Pole. This law also implies that there are no magnetic monopoles. See Equation 5.5. and Figure 5.5.

The third equation, based on Faraday's work, states that a changing magnetic flux generates an electric field. This phenomenon is employed in electric generators and transformers. See Equation 5.6 and Figure 5.6.

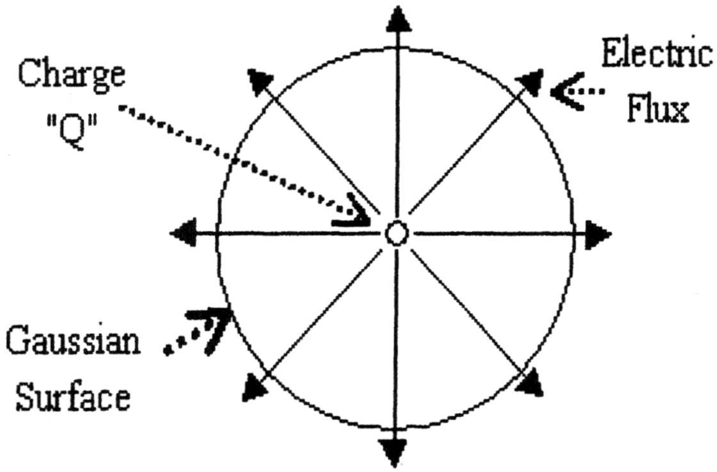

Figure 5.4 **Electric Flux From a Charge**

∂ **B**.d**A** = 0

Equation 5.5 **Maxwell's Second Equation**

∂ **E**.d**s** = -d/dt(∂ **B**.d**A**)

Equation 5.6 **Maxwell's Third Equation**

The fourth, based on Faraday's work, states that magnetic flux density is proportional to the rate change of electric field. See Equation 5.7 and Figure 5.7

The key to understanding EMC issues is to find, understand and apply these four relationships to every EMC issue. This will allow insight into the unifying physics of the problem and make it possible to understand and solve.

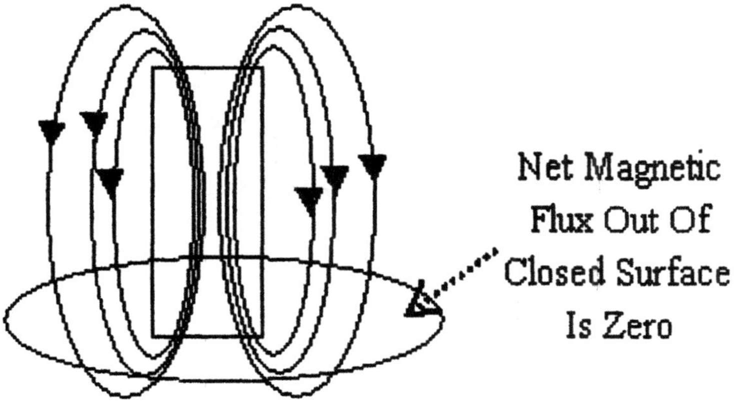

Figure 5.5 **The Net Magnetic Flux Out of Any Closed Surface is Zero**

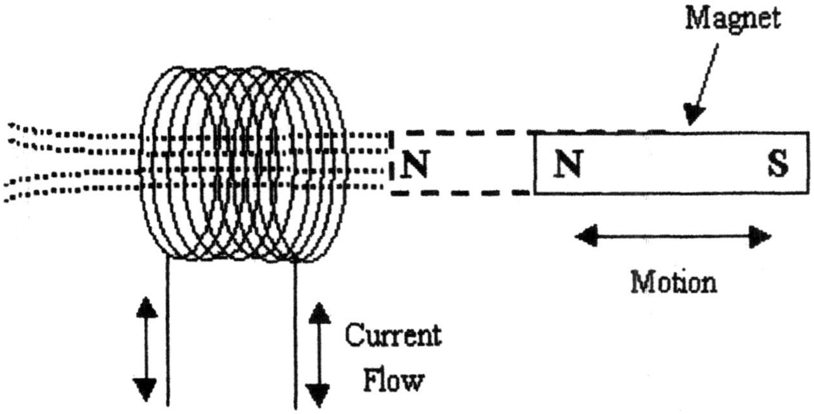

Figure 5.6 **A Changing Magnetic Field Produces Current Flow in a Nearby Coil**

∂ B.ds = m$_0$.(I + d/dt (ε_0. ∂ E.dA))

Equation 5.7 **Maxwell's Fourth Equation**

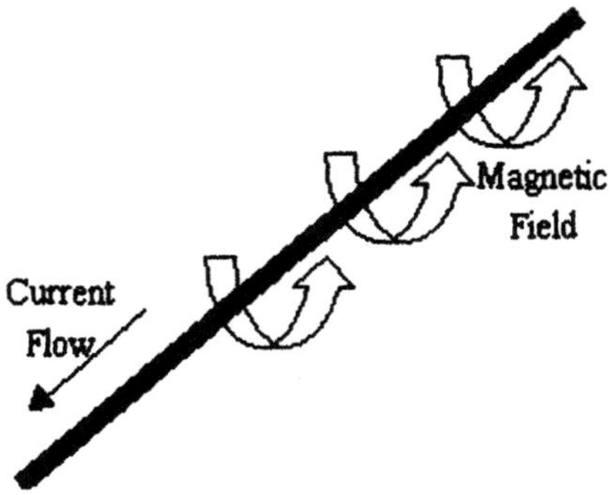

Figure 5.7 **The Magnetic Field Around a Wire is Proportional to the Current Flowing in the Wire**

5.5 REGIONS AROUND A SOURCE:

There are two additional concepts to understand describing the field regions surrounding an antenna radiating power, as shown in Figure 5.8.a and 5.8.b.

5.5.1 Far-Field

The region extending farther than two wavelengths away from the source is called the "far-field." The equations described here are not valid in the near-field or transition zone. They are only valid at distances further than $\lambda/2\pi$ or $2D^2/\lambda$, whichever is greater. In the far-field, E, H, and power density are related by the equations: E = H x 377 and P_d = E x H. Combining these two equations together we get:

$P_d = H^2 \times 377$ and $P_d = E^2 \div 377$

Where

H^2 = Power density in Watts per square meter (one W/m^2 is equal to 0.1 mW/cm^2),
E^2 = the square of the value of the electric field in volts squared per meter squared.

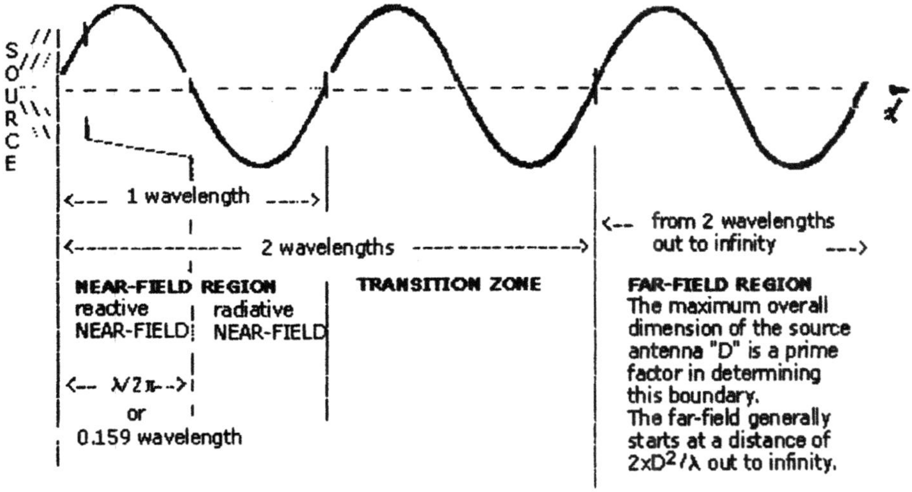

Figure 5.8a. **Antenna Field regions**

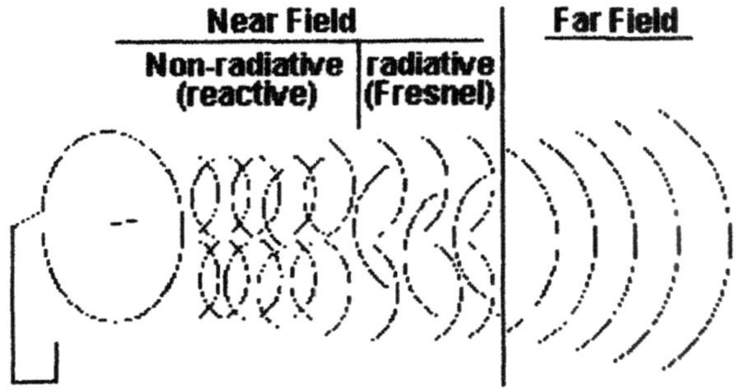

Figure 5.8b. **Antenna Field Regions**

Far-Field Distance:
Consider a small radiating element 30 cm long
For a radiator, at 100 MHz, $\lambda/2\pi = 3m/2\pi = 48$ cm
and $2D^2/\lambda$ is $2\,(30cm)^2/3m = 16.2$ cm
So the far field begins at the greatest of 16.2 or 48 cm, or 48cm.

If the radiating element were 3m long,
$\lambda/2\pi = 3m/2\pi = 48$ cm, as before
and $2D^2/\lambda$ is $2\,(3m)^2/3m = 3$ m
So the far field begins at the greater of 48 cm or 3m, or at 3m.

The above equations show that in the far-field, all we really need to measure is the E field (E^2 on some measurement instruments.) From this measurement, the power density and value of the H field can be calculated. We would prefer to do our EMC analysis and data collection in the far field region of the antenna. The measurements we take will be less sensitive to antenna position than what would be the case in the near field. For many EMC problems, especially with immunity due to external sources, this also represents the "real world" conditions.

5.5.2 Transition Zone

The region between the near-field and the far-field is called the "Transition Zone". It exhibits some of the characteristics of both the near-field and the far-field. Here it may not always be necessary to measure both E and H to obtain a good approximation of the EM field, but several measurements are needed to characterize the field.

5.5.3 Near-Field

The region located less than one wavelength from the source is called the "near-field." Here, the relationship between E and H becomes very complex, and measurement of both E and H are required to determine the power density. Also, unlike the far-field where EM waves are usually characterized by a single polarization type (horizontal, vertical, circular, or elliptical), all four polarization types can be present in the near-field. The relationship between E and H is complicated in the near-field.

The near-field is further divided into the "reactive" near-field and the "radiative" near-field.

In the reactive near-field (very close to the antenna), the relationship between the strengths of the E and H fields is too complex to predict. Either field component (E or H) may dominate at one point, and the other way dominate at a point only a short distance away. This makes it extremely difficult to find the true power density. Not only would E and H both have to be measured, and a new term called the phase relationship between E and H would be needed.

In addition, the reactive near-field region has another characteristic. In this region, not only is the EM wave being radiated outward into space there is also a "reactive" component to the EM field. Near the antenna, energy of an unknown amount is held back and is stored near the antenna surface. This reactive component can be the source of confusion and error in attempting measurements in this region. In other regions the power density is inversely proportional to the square of the distance from the antenna. In the vicinity very close to the antenna, the energy level can rise dramatically with only a small movement toward the antenna.

In EMC terminology, electric field sources (dipole-type antennas, for example) are referred to as high impedance sources, and magnetic field sources (loop antennas, for example) are referred to as low impedance sources. This makes sense as it implies that current causes magnetic fields, which implies low impedance.

Why is this important to EMC? The reason is that we may be required on occasion to take measurements in the near-field, for example, when measuring emissions from devices and components installed in a vehicle.

Where does energy go when it radiates from an antenna? Since current always must return to its source, we can think of the current as flowing through the parasitic capacitance in free space as shown in example of the quarter wave vertical antenna and the dipole antenna in Figure 5.9. This means that if a transmitting antenna is mounted on a vehicle, the RF current will flow through the vehicle body to create a return path to the antenna feed point. This current flow then induces an electric field, which generates a corresponding magnetic field, and since this process repeats over and over again this causes the propagation of radiated energy.

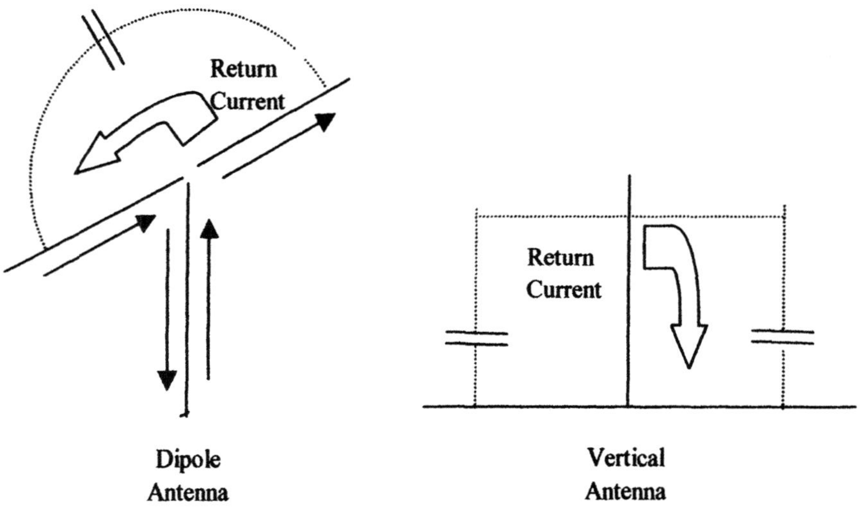

Figure 5.9. **Antenna Current Must Return to the Source**

5.6 POLARIZATION

Polarization is an important concept in making electromagnetic measurements. It explains why "walkie-talkie" antennas need to be oriented in the same polarization (usually vertical) to get best reception and why RF survey probes must be used in successively different orientations when we measure an RF field.

Polarization of a radiated wave is "That property of a radiated electromagnetic wave describing the time-varying direction and amplitude of the electric field vector: Specifically, the figure traced as a function of time by the extremity of the vector at a fixed location in space, as observed along the direction of propagation." While the above definition can be confusing, the following discussion can decrease that confusion.

Radiated EM waves traveling through space have a property called polarization. It affects the compatibility of waves and certain types of antennas. There are several parameters, which cause some antennas to accept one wave and reject others:

- The physical size of an antenna (the antenna aperture) influences what wavelength (or what frequency) will be efficiently radiated or received by that antenna.
- The shape of the antenna helps determine the directivity of an antenna. Directivity involves the compass direction at which an antenna radiates or receives EM waves.
- The property of polarization describes the angular positioning of the EM field vector.

All three of these properties (physical size, directivity, and polarization) are separate and distinct properties.

There are several types of polarization: elliptical, circular, and linear. The polarization type is determined by the angular pointing of the electric field vector.

To determine the polarization type, imagine observing the tip of the time-varying electric field vector from a fixed point in space along the direction of the wave's propagation. The image traced by this vector tip is elliptical, but commonly the ellipse degenerates to a circle or a straight line.

Figure 5.10 may help in visualizing polarization of EM waves:

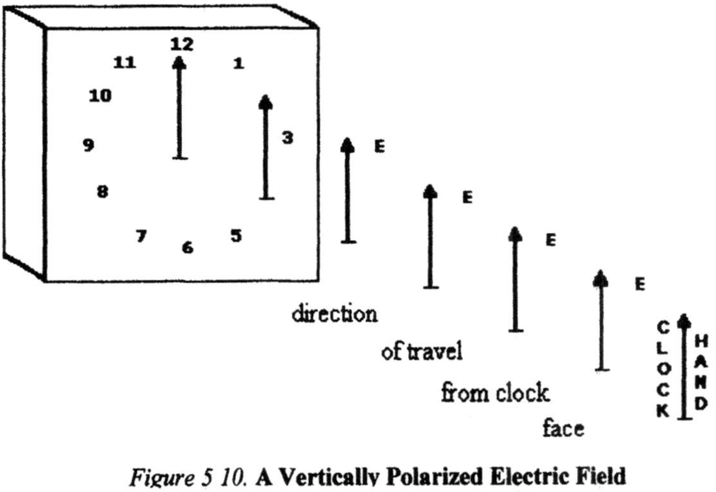

Figure 5 10. **A Vertically Polarized Electric Field**

As shown in the figure, imagine a clock face with one hand pointing to the 12 o'clock position. Let the hand move out away from the clock face. Immediately after the first hand leaves the clock face, let a second one replace it on the clock and also move out away from the clock face. Repeat this again and again until a steady stream of clock hands are flowing away from the clock, all pointing upward at the same angle. The clock hands are representative of the vector of a vertically polarized electric field as it moves out from the source.

EM waves vary in amplitude during the period of one cycle. This variation repeats over and over again for each cycle of the wave as it is radiated. Let us move ourselves from our viewing position to a new position, one looking at the side of the clock. If we allow each subsequent clock hand (E field vector) to vary in size (amplitude) from the previous one, we get a side view as shown in Figure 5.11.

If we now return to our original viewing position, the vertically pointing clock hand example is comparable to a vertically polarized electric field. If someone were to reach out to catch one of the clock hands, he can catch it only if his hand is positioned at the same angle (polarization) as the clock hand coming broadside at him. Remember, the arrows are not pointed at him, but are pointing up and down. If his hand is turned sideways, different from the angle of the clock hands, he would not be able to catch any.

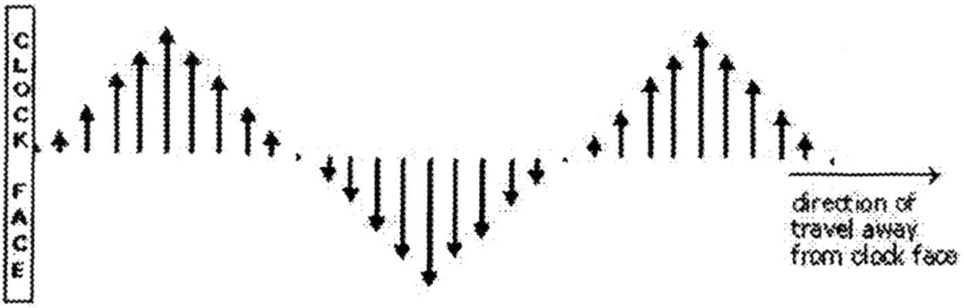

Figure 5.11. **The Wave in Figure 5.6 Varying Sinusoidally in Amplitude**

If his hand is oriented vertically, he can catch a vertical arrow, but not a horizontal arrow, and conversely.

Just as the pointing of the electric field vector determines the EM field's polarization, the H field is also dependent on the E field's vector. To help see this, add another hand to the clock face so there are now two hands on the clock, perpendicular to each other, as in Figure 5.12.a and 5.12.b.

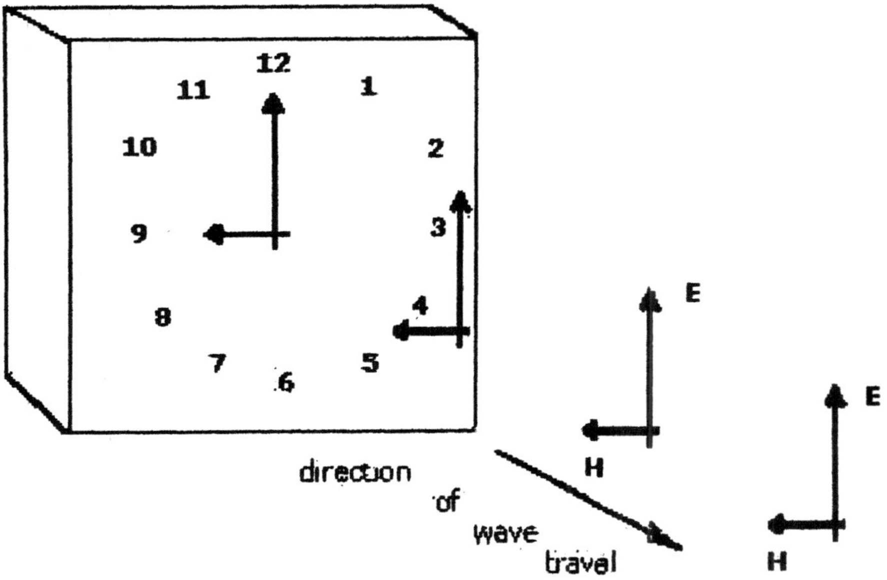

Figure 5.12a **E and H Field for a Vertically Polarized Wave**

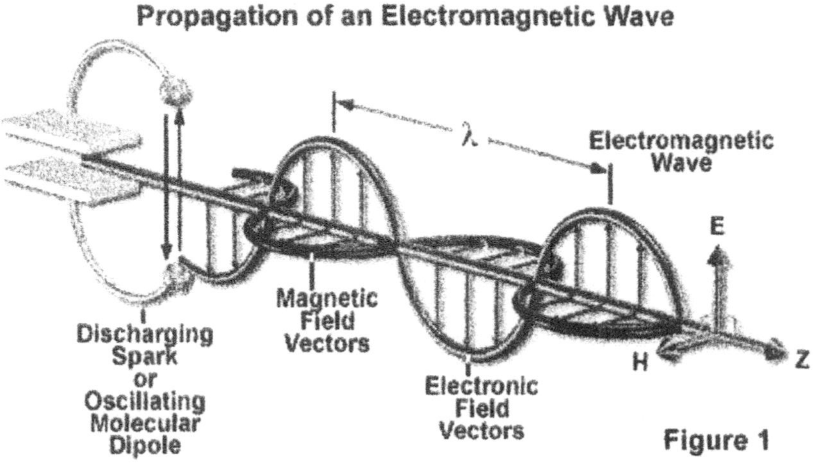

Figure 5.12b **E And H Field for a Vertically Polarized Wave**

The two clock hands in Figure 5.12a represent the E field and the H field vectors. If one hand is positioned at 12 o'clock and is called the E field vector, the other hand at the 9 o'clock position would be the H field vector. The two vectors are at right angles to each other. Since polarization is determined by the E field vector, the illustrated polarization is vertical. (If the E field vector were pointed to the 3 o'clock position, the polarization would be horizontal. If the E field vector is rotating, the polarization is circular or elliptical.) Unlike ordinary clocks, the clock in Figure XY requires the two hands always be locked together at a 90-degree angle. The H field vector (illustrated by the 9 o'clock hand) is always perpendicular to the E field vector. The hands can be pointed (tilted) in any direction, but must always be perpendicular to each other. The "right hand rule" illustrates that the direction of propagation is determined by rotating the electric field vector into the magnetic field vector, similar to wrapping one's hand around a wire. The direction of propagation is then determined by the direction of the thumb, as in Figure 5.13. If an antenna's orientation is tilted sideways at an angle, the polarization of the transmitted EM field would tilt by the same angle, but the E and H fields still remain perpendicular to each other.

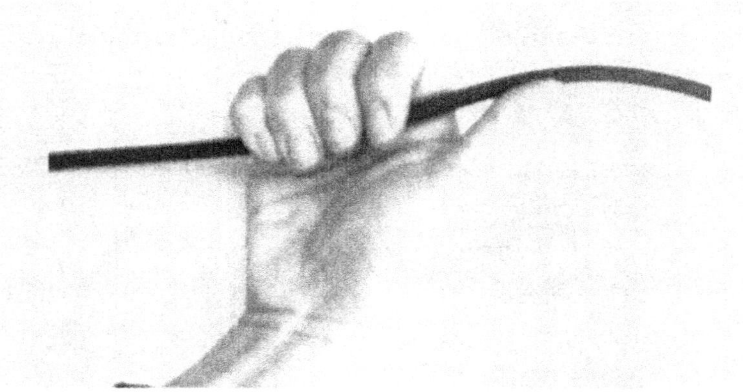

Figure 5.13. **Right Hand Rule for Direction of Propagation**

As explained in the above illustration, the polarization of the EM field is referenced to the E field, with the associated H field at a right angle to the E field. The transmitting antenna determines the polarization angle of the electric field radiated from it. Polarization effects are most readily observed at frequencies above 30 MHz. A low VHF band (30 - 50 MHz) radio antenna pointed straight up would radiate a vertically polarized wave, and a horizontal "dipole" (similar to roof-mounted TV antennas) would radiate a horizontally polarized wave. The best reception is obtained when the receiving antenna is polarized (tilted) to match the polarization of the transmitting antenna. That is why mast antennas used for mobile radio all point the same way, vertical, since it is easier to mount a vertical antenna on a vehicle than a horizontal one.

The following experiment visually demonstrates polarization and the importance of matching the polarization between a source antenna and the receiving antenna:

1. Take two pairs of "polarized" sun glasses. They must be polarized.
2. Use one pair to filter the light coming from a flashlight.
3. Wear the other pair.
4. Now tilt your head 90 degrees sideways and notice that at a particular head angle one eye receives the transmitted polarized light and the other receives none.
5. Rotate the polarized sun glasses positioned at the light source.
6. Now do the head tilts again and notice the polarization angle has changed by the amount rotated in step (5).

NOTE: On a sunny day items on an automobile dashboard can be seen reflected in the windshield, but the images are much less visible if you are wearing polarized sun glasses (provided the windshield is tinted). If you tilt your head while wearing the polarized sun glasses the image reflection will appear and disappear at 90 degree angles.

When performing an electromagnetic radiation field survey, the instrument's probe is usually an isotropic receiving antenna. An isotropic[1] probe receives electromagnetic signals regardless of polarization or direction of travel. Such probes are constructed using several antennas arranged in three separate and perpendicular planes. An isotropic probe is designed to give the same reading, no matter which way the isotropic probe is oriented in the EM field.

The clock and arrow illustration was designed to help the reader understand the difficult concept of polarization. Electromagnetic waves do not actually transfer energy as "arrows" or "small packets" of energy. It would be a mistake to think of RF energy transfer as anything other than a wave whose energy is transferred by the time variation of electromagnetic fields.

---

[1] Actually, "polarization insensitive," but commonly referred to as "isotropic."

5.6.1 Magnetic Field Emissions

To control magnetic field emissions, the primary goal is controlling loop areas. For instance, when the input and output power and signal leads are twisted together, and the traces and wiring within the equipment are routed close to each of their current returns, it minimize the loop area. Minimizing loop area provides optimum cancellation of magnetic and reduces overall radiated magnetic fields.

The most common source of magnetic field emissions in a switched mode power supply is the "high amp-turns" (many turns carrying high current) components or magnetics. One method to reduce the leakage flux from magnetic core gap transformers is adding a shorted turn for the leakage flux. This turn goes around the entire magnetic device and causes an opposing current to the leakage flux. When the flux couples to the shorted turn, a current is induced in the direction such that the resulting flux opposes the incident flux, which changes the pattern of the radiation. This change in the radiation pattern reduces the area radiated by the magnetics.

For shielding low-frequency magnetic fields, loss due to reflections is the primary field-shielding mechanism. The incident magnetic field induces a surface current in the shielding material, which in turn re-radiates. The re-radiated field is (almost) equal in magnitude and opposite in phase to the incident field. If a discontinuity exists, the currents are disrupted and the re-radiated field will not cancel with the incident field. This disruption in current cancellation will degrade the shielding.

5.6.2 Modeling/Prediction Techniques

While prediction methods are available for predicting RE, there are complicated programs which require tedious circuitry input parameters. More general calculation routines are available in which the calculations are done with pencil and paper or computer spread sheet. Table 5.4 gives an example of the general type calculation and presents a series of columns which are added to determine the predicted RE from a given component. Basically, the Fourier transform is computed for a given signal. The amplitude (dB μV) is placed in column 1. A correction factor is added to account for the conversion of the conducted data to free space. This factor is assumed to be –34 dB for a perfect $\lambda/4$ The frequency, f_3, is computed to give a factor that takes into account the actual length of the cables. The calculation is shown in Table 5.4 and corrects the $\lambda/4$ assumption. Finally, the number of leads from the signal is accounted for. The measurement distance is also be factored in; for most test applications it is 1 m and requires no correction factors. The resulting factor is compared to the specification at that frequency to determine the dB of attenuation needed.

The attenuation is obtained by using metal equipment housing, shielding, twisting, etc. This prediction method is reasonable for checking the most likely culprits in the circuits (diodes and transistors that switch high levels of current).

Table 5.4 **RE prediction Analysis**

Frequency	(1) Cn	(2) Voltage to E Field	(3) Antenna Factor	(4) Number of Leads	(5) Distance	(6) Result
20 MHz	130 dBμV	-34	-2.7	6	0	99.3 dBμV

(1) Frequency domain amplitudes - Fourier transform
(2) Voltage to field intensity level 1 meter away from conductor = –34 dB
(3) -10 log (f_3/f_x) where $f3$ □3 $(10)^8/4L$, L=wire length in meters when fx/f_3 the correction factor = 0 dB.
(4) + 10 log N, where N = number of leads
(5) - 20 log D, where D = test distance in meters
(6) Result in dBμV/m, sum (1) through (5).

The above table is shown as a flow chart in Figure 5.14.

Modeling can be used to help formulate an EMC test plan, by predicting which frequencies couple to wiring harnesses and modules. It can be used to decide whether additional isolation is needed or not, thereby letting the design engineer know whether to budget for shielding components or twisting wiring harnesses. It can even be used to help orient modules (for example, Remote Key Locks or Remote Start) with built-in antennas for maximum coupling to external sources of energy, when desired.

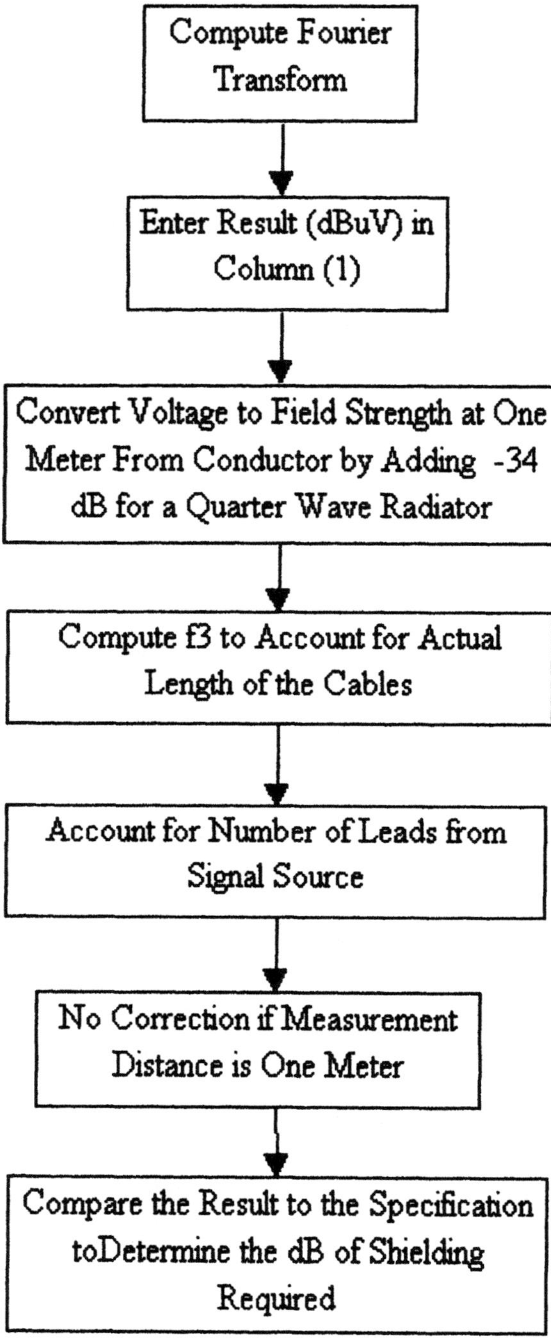

Figure 5.14 **RF Prediction Analysis**

Chapter 6

EMC Testing

6.1 EMC DISCIPLINES

Why is understanding all the EMC disciplines important? The typical EMC problems can involve any combination of the disciplines, as well as combinations of frequencies, dimensions of components, wiring harnesses, and assemblies. This can make problem solving a challenge, requiring seemingly contradictory approaches for solutions at different frequencies or under different conditions.

EMC is divided into three disciplines based on coupling mechanisms. The first is the radiated path, the second is the conducted path, and the third is a combination of the two mechanisms (sometimes encountered with electrostatic discharge [ESD]). Within each area are two additional disciplines - the first emissions, and the second immunity. These are shown in Figure 6.1.

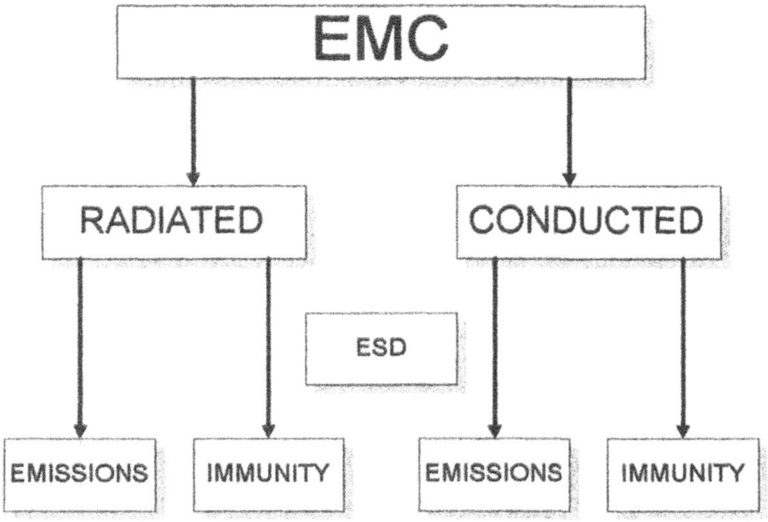

Figure 6. 1. **EMC Disciplines**

On the left side of the figure, radiated phenomena are divided into radiated emissions (RE), and radiated immunity (RI).

On the right hand side of the figure, conducted phenomena are also divided into emissions (CE) and immunity (CI).

In the center of the figure is ESD, which may consist of a combination of radiated and conducted phenomena.

Let's examine an example of each one of these disciplines:

If we have a brush-type DC motor, this may affect the reception on a nearby AM radio (you may have heard this in your car). This is an example of "emissions" from the motor and/or its wiring harness, and the "immunity" of the radio not being compatible. Either the emissions of the source must be reduced, or the immunity of the victim must be increased for the radio reception to be free of interference.

The mechanism of coupling in this instance is from the source (motor brushes), radiating via the twelve-volt power wiring harness, which acts as a transmitting antenna. See Figure 6.2.

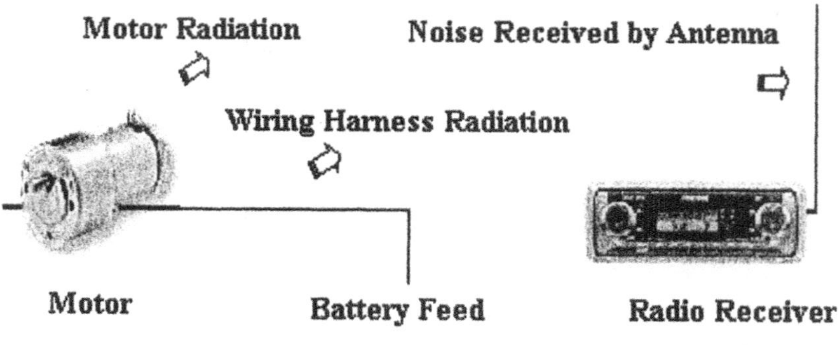

Figure 6. 2. **Radiated Coupling**

Radiated emissions (RE) and radiated immunity (RI) are expressed in terms of microvolts per meter or volts per meter, or their dB equivalent. This means that a 1-meter-long piece of wire going from a 10 V/m point in space to a 20 V/m point would have a voltage of 10 volts across it. For magnetic field coupling, the units would be microamps per meter or Amperes per meter.

How real are EMC problems as a result of radiated emissions? There has been a documented case of an automobile that was sensitive to radiated path energy. When this particular make of automobile was approaching a certain portion of the highway, near a broadcast transmitter, it would experience problems with its electronic control system. The solution was to erect a screen alongside the highway to prevent the transmitter energy from being radiated onto the roadway. This is an example of a problem with the radiated immunity of the vehicle being incompatible with the electromagnetic environment in which the vehicle must operate.

Another phenomenon sometimes observed on car radios is "alternator whine." This is an example of "conducted emissions" as shown in Figure 6.3. The audible ripple frequency voltage travels via the shared power wiring to the power connector on the radio. This is an example of the "conducted emissions" from the vehicle, and the "conducted immunity" of the radio not being compatible. Conducted coupling paths are often much more efficient than radiated coupling paths, which means it takes much less energy to cause similar problems.

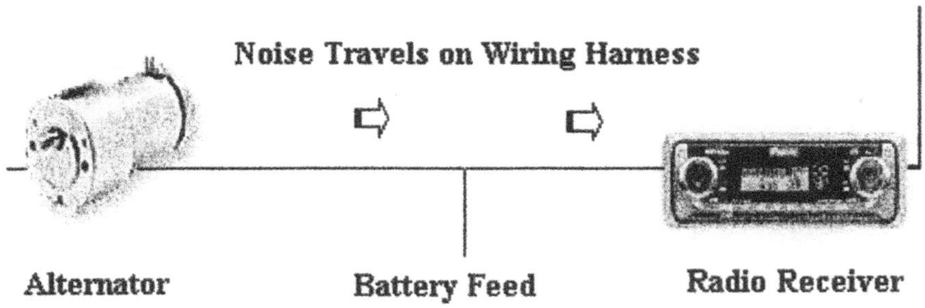

Figure 6. 3. **Conducted Coupling**

Conducted emissions (CE) and conducted immunity (CI) are likewise expressed in terms of volts, or amps, or their dB equivalent.

Now that we've reviewed the basics of EMC and their importance to the automotive industry, we can focus on a particular aspect of EMC disciplines -- that of the radiated path. In order to discuss the characteristics of the radiated path, it is important to review the fundamentals with regard to electric and magnetic field waves and their propagation. Key to this is understanding what is called the "right hand rule ". This is shown in Figure 6.4. If your thumb represents the direction of (conventional) current (positive charges), then as your fingers surround the conductor they will indicate the direction of the magnetic field surrounding the conductor. The electric field radiates outward perpendicular from the conductor, and the magnetic field encircles the conductor.

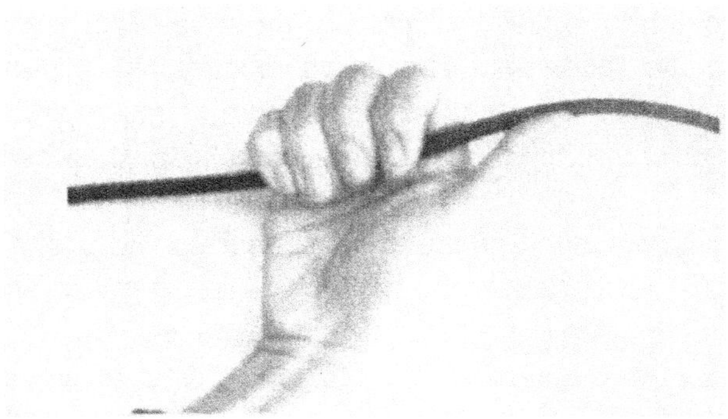

Figure 6. 4. **Right Hand Rule**

6.2 RADIATED EMISSIONS DIAGNOSTICS

RE are caused by the flow of current on conducting surfaces. These currents include intentional signals on wires in cables, unintentional noise flowing on the same cables, and currents flowing on equipment enclosure surface. The following techniques assist in determining the source of radiated emissions and in curing the problems.

6.2.1 Low-Frequency Specification

If emission frequency is such that dimensions of the test setup are small relative to a wavelength, then equipment-connected cables are the primary suspects. Efficiency of a circuit as a radiator is proportional to its length when the length is short with respect to a wavelength. Even though the intentional signal is a baud rate (10 kHz or slower), MHz signals may parasitically couple to the wire or the outside of a cable shield. Using a current probe with suitable bandwidth is indicated here. Various models of current probes (Figure 6.5) are available, ranging from 20 Hz to 1 GHz.

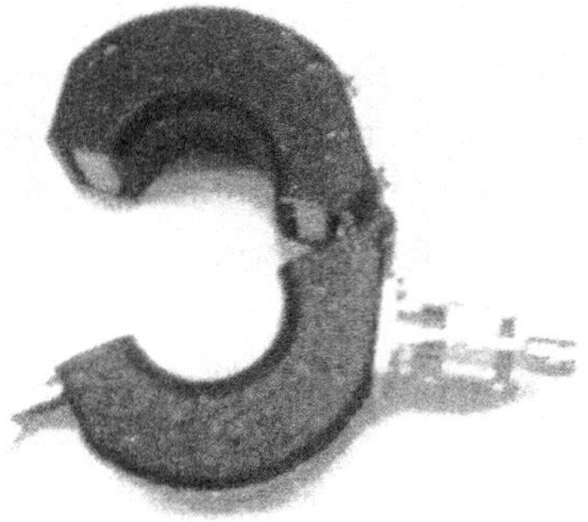

Figure 6. 5. **Current Probe**

Emissions at 400 MHz and above are more likely to emanate from the equipment enclosure itself. If significant CE are found at a frequency corresponding to out-of-specification RE, then mitigating steps in the next two sections are worth-while.

6.2.2 Bulk Current Injection

BCI is a lumped element model of field-to-wire coupling. As such, it is most applicable at low frequencies, in which the CUT (Component Under Test) is small relative to wavelength. Figure 6.6 stops at 400 MHz, since fields above this frequency are just as likely to penetrate the equipment enclosure as to couple to the cables and since serious concern exists with the validity of the test at higher frequencies. An injection clamp similar to that used in CE measurements injects currents onto the CUT. Figure 6.6 shows the induced current expected as a function of frequency due to a 1-V/m field impinging on a 2-m CUT. Computation from Faraday's law and typical cable installation geometries yields the result that 1.5 mA of current flows on a cable in response to 1 V/m of incident field intensity at frequencies in which the cable is at least one-half wavelength long. At lower frequencies, the induced current drops at 20 dB per decade. If field intensity is different from 1 V/m, the dB mA and dBm curves are adjusted as 20•log (actual field intensity in V/m). If the cable is longer than 2 m, the low-frequency breakpoint is extended in direct ratio to the length extension. If the BCI clamp insertion loss differs from that plotted, the dBm curve shifts accordingly. Figure 6.6 shows why the test is a good precompliance tool: power requirements at the clamp are compatible with a signal generator output and no expensive amplifier is necessary. Figure 6.6 assumes a clamp covering 2 to 400 MHz. Injected current at lower frequencies is so low that only tunable RF electronics operating in-band to the susceptibility signal are expected to respond. A typical BCI clamp is shown in Figure 6.7. If this test is performed at an EMI test facility, a refinement is available: use a current measurement probe and spectrum analyzer for over-current control. Regardless of predicted clamp drive power from Figure 6.6, actual injected current should not exceed the dB mA curve by more than 6dB.

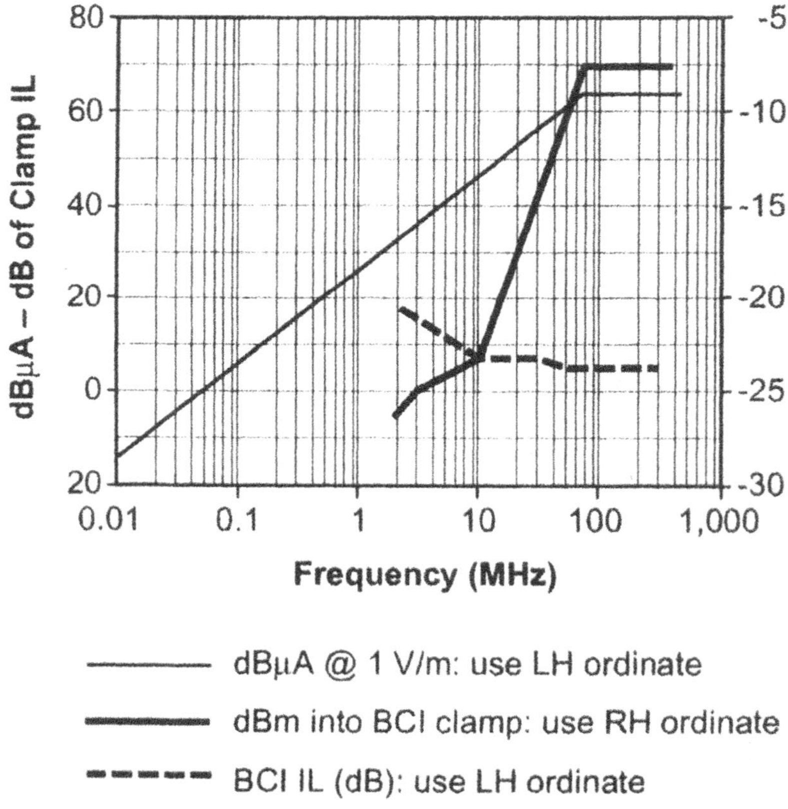

Figure 6. 6. **Converting a One-Volt-Per-Meter Field to Bulk Current**

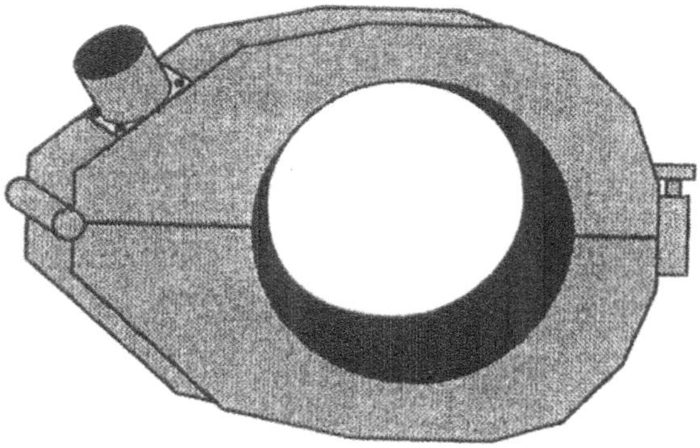

Figure 6. 7. **Typical Current Injection Clamp**

6.3 HOW A SWITCHING TRANSIENT OCCURS

Figure 6.8 shows the elements of a power distribution system - a power source, distribution wiring, and load. The power source is simplified to an ideal voltage source in series with a resistive and/or inductive impedance. The distribution wiring contributes both resistance and inductance. The load, at turn on or turn off, provides a rapid change of current through the power source and wiring impedance. This simple model ignores any capacitive effects other than the load. Source parallel capacitance (especially in a dc supply) contributes to low source impedance which is easily modeled in the transient case by using a smaller series source impedance. Line-to-line or line-to-return wiring capacitance is easily accounted for by modeling the distribution wiring as an inductance bypassed by a resistor; i.e., a lumped element model of a transmission line (LISN). Figure 6.8 shows a model for both calculating and measuring switching transients. In Figure 6.9, the LISN models the distribution wiring impedance.

The 5.0- μH, 50-Ω LISN has been selected to serve as worst-case model of vehicle wiring. There is some intuitive rationale for the selection. Consider that the inductance of a wire above a signal return plane is roughly one microhenry per meter (for typical geometries). Five micro-henries account for a wire length of 5.0 m, which is certainly a reasonable worst case for this type power distribution (only feasible in metallic vehicles). A two-wire line has an inductance about one-tenth of the wire-above-signal return. There are also LISNs with 50 μH inductance which represents about 500 m of wiring in a commercial setting (also a worst-case model).

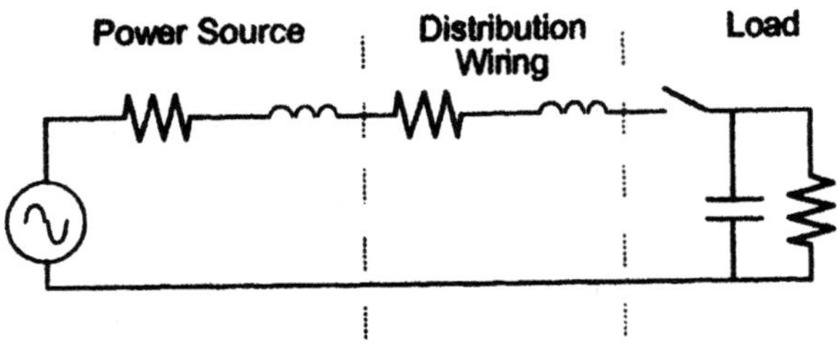

Figure 6.8. **Model of Electrical Power Distribution System**

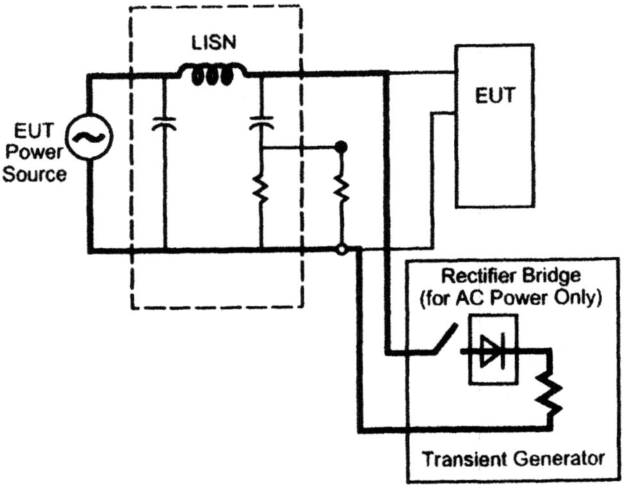

Figure 6.9. **Spike Generator (Heavy Lines Show Flow of High Current to Spike-Generating Load)**

The transient generating mechanism is the switching on/off of a heavy power bus load. If the immunity of the EUT to other power bus load induced spikes is to be assessed, a heavy load must be switched from the LISN while the EUT is in steady-state operation and shares the same LISN as a power bus source impedance. The LISN models the common impedance to the EUT and switched load.

6.4 TEST METHODS

Basic EMC Test Hardware

The objective of EMC testing is to demonstrate conformance to requirements. Failure to meet requirements may indicate a need for redesign; and, typically, further analysis is first performed to determine if the particular failure is likely to cause an EMC problem. For example, an equipment emission that exceeds the radiated emission (RE) limit by 20dB at 100 kHz may not be serious if the overall system for which the equipment is destined does not utilize the spectrum below 2 MHz.

We will discuss some of the basic instrumentation utilized when performing EMC tests on automobiles. Many of these techniques and instruments are also applicable to EMC in other areas, such as home electronics or military hardware. This section will focus on three main areas, equipment, transmission lines and cables, and EMC measurements.

6.4.1 EMC Instrumentation

The majority of EMC signal generating instrumentation has a source impedance of 50 ohms, as shown in Figure 6.10.

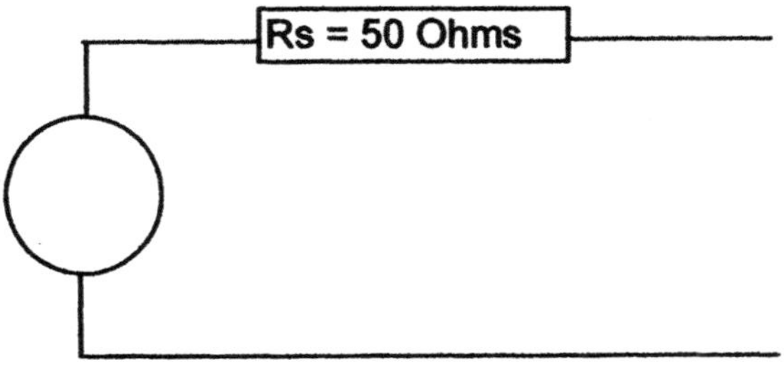

Figure 6.10. **Internal Impedance of RF Signal Source**

Most EMC measurement instrumentation has an input impedance of 50 ohms (Figure 6.11). There are some exceptions to this; volt meters and oscilloscopes may have higher impedance so as to not load down the circuit being monitored during testing.

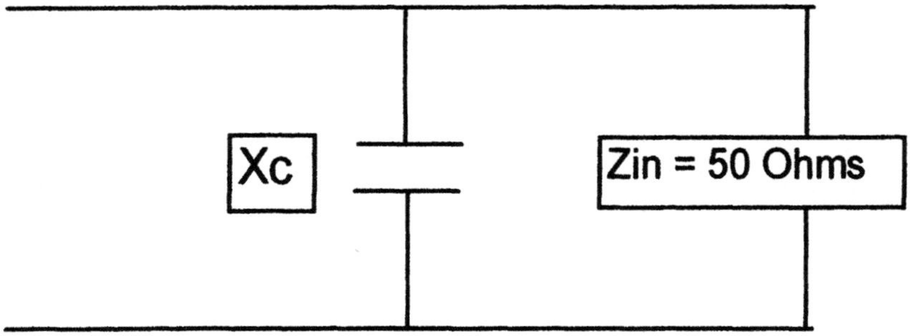

Figure 6.11. **Impedance of Most RF Measurement Instrumentation**

Why is 50-ohm impedance so common on test equipment? The reason is that it is intended to match the impedance of the majority of coaxial cable that used in most laboratories. There can be complications associated with using impedances other than 50 ohms. The measurement will become frequency dependent, based on this mismatch of cable, source, and load impedances.

6.4.2 Amplifiers

One of the most important components of the RF immunity test system is the power amplifier. Amplifiers are used both for radiated testing using antennas or TEM cells, and for conducted testing using current injection probes or artificial networks. The transducer used has a major effect on the specification of the amplifier, and the two could be considered a system. Amplifiers are often housed in a special room so that adequate cooling can be provided, and so that access can be limited to authorized personnel.

Power Output and Bandwidth

Two specifications, maximum power output and the bandwidth over which this output is sustained, are determined by the required capability of the test facility. The bandwidth and power requirements depend on which standards the equipment under test (EUT) will be tested to, and the test configuration.

Frequency Range

European Directive 72/245/EEC covers frequencies from 20 MHz to 1000 MHz. Revisions are being considered to extend radiated immunity testing up to 2 GHz. Many manufacturers test at frequencies outside this range, as well as perform conducted immunity testing, to satisfy their own internal requirements. Any overlap between conducted and radiated testing frequencies depends on internal standards.

A fundamental trade-off exists between power and bandwidth. An amplifier with a few Watts output can be made to cover several decades bandwidth (e.g., 100 kHz to 1 GHz). However, as the power capability increases, restrictions in output stage design mean that less bandwidth can be obtained. For this reason, check that the rated output power is available over the entire specified bandwidth. The bandwidth may, for instance, be specified as a 3-dB bandwidth, which is likely to mean that only half the rated power is available at the band edges.

However, the full power capability may not be needed over the entire frequency range because of transmit antenna characteristics. It is reasonable to use amplifiers of different power ratings to cover different frequency ranges. It would, of course, be possible to use one amplifier for conducted testing and a different one for radiated testing, and this is possible because the power levels and frequency ranges are different. The need to change amplifiers or antennas manually can be time consuming and decrease the amount of actual test time in the facility.

Test Level

Assuming that one is working primarily to 72/245/EEC, a test level of 30 V/m with 80 percent modulation is required. Considerations should be given system losses in cables, connectors, couplers, and switches. Losses increase with frequency but may be offset by increasing antenna gain at the higher frequencies.

Amplifier power requirements may depend upon:

- In a test chamber, the antenna gain and the intended test distance.
- For conducted tests, the factors of the various transducers used.

Table 6.1 summarizes the calculated power requirements for radiated testing with a biconical/log antenna, assuming +5.2 dB for 80 percent amplitude modulation, and +3 dB for field uniformity.

Table 6.1. **Calculated Power (Watts) for Radiated Testing.**

	1-meter Distance			3-meter Distance		
	3 V/m	10 V/m	30 V/m	3 V/m	10 V/m	30 V/m
27 MHz	25.4	282	254	228.8	2542	22880
80 MHz	1.29	14.3	129	11.59	128	1159
200 MHz	0.29	3.23	29	2.61	29.0	261
1 GHz	0.33	3.64	33	2.95	32.7	295

Conducted Testing

The commonly used bulk current injection (BCI) method requires more power than some other methods since the injection probe is fairly lossy at the extremes of its frequency range. A 100-Watt amplifier is frequently used.

Modulation

Another factor which affects the power requirement is the need for amplitude modulation. The test level is defined in terms of an unmodulated signal; consequently, the field calibration method uses an unmodulated signal. When the test is actually run, modulation is applied. The default modulation is a 1 kHz sinusoid at 80% modulation. The relationship between modulation depth and the amplitude envelope of the signal is shown in Figure 6.12. The peak signal power is increased by 5.2 dB, so the amplitude of the immunity field is reduced by this same amount. This adjusts the peak signal strength with modulation to same peak field strength as without the modulation.

VSWR (voltage standing wave ratio) is a measure of the match to a resistive 50 ohms that is presented at the terminals of an amplifier or a transducer. Unless the impedance is exactly 50 ohms, some power is reflected from the terminals and travels back down the feedline.

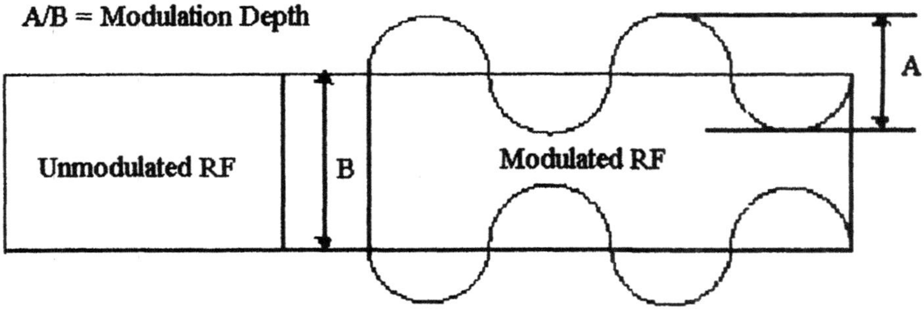

Figure 6.12. **Relationship between Modulation Depth and Amplitude Envelope**

VSWR Tolerance

A VSWR of 1:1 implies a perfect match. An open or short circuit implies an infinite VSWR. A biconical antenna can show a high VSWR (30:1) at 30 MHz. Even at the higher frequencies, coupling with the screened room and the EUT can increase the VSWR. If coupled to a current probe for BCI injection, VSWRs greater than 60:1 can result. Under these conditions, much of the applied RF power is reflected back to the amplifier output.

A power amplifier for EMC immunity testing must be able to deliver as much of its rated power into mismatched loads as possible.

Class A or AB?

The definitions of Classes A and AB, originally developed for tube-type amplifiers, are not strictly applicable to solid-state amplifiers. An ideal Class A amplifier has an operating current that does not vary with output power. It can be either single-ended or push-pull. In push-pull operation, each valve still conducts through 360°. The advantage of push-pull is that the non-linearity of the tubes should cancel. Class A has the lowest efficiency, and therefore the highest cost, but, as it is designed to dissipate all the DC power supplied to it when no input signal is present, it can also tolerate total reflection of the output power.

Class A is also the only class of amplifier that will operate with very fast (e.g., pulse) modulation signals. In the other classes the amplifier operating current varies with the modulating signal, and the intrinsic parasitics of the amplifier circuits will not allow rapid variation of the operating current.

Class AB allows for a slight overlap at the crossover point. Each tube conducts over more than 180°, thus reducing crossover distortion. A distinction between Classes AB1 and AB2 is the presence or absence of positive grid current. Class AB has better efficiency than Class A, but the amplifier needs to be protected against excess reflected power.

These classifications must be applied with caution to solid state amplifiers. The great majority of designs use field effect rather than bipolar transistors. A typical FET amplifier will operate in Class A, but the quiescent current can be lowered in order to lower dissipation when operating at low power. This does not reduce ruggedness at high power levels, provided that the cooling provision is not also reduced, so Class AB can be more reliable than Class A. Many MOSFET amplifiers operate in a mode between Class A and Class AB, whereas some GaAsFET amplifiers operate in Class A. These operating modes are considered to offer the best reliability, performance and cost tradeoff.

Linearity

When a power amplifier is used for EMC immunity testing with a modulated signal source, the amplifier must remain linear over its entire frequency range up to its maximum power. If it does not, there are two consequences:

- The peak of the modulated RF envelope will be flattened. This creates harmonics of the modulation frequency (1 kHz.). It will also reduce the actual peak applied field strength.
- Harmonics of the carrier frequency will be generated. If the non-linearity is severe, the harmonic amplitudes can approach the fundamental, and this can result in susceptibilities being observed at the wrong frequency. It will also reduce the applied field strength at the carrier frequency. New combiner technology used in the higher power amplifiers results in exceptionally low harmonic performance, 30 dB or better. IEC 1000-4-6 specifies that harmonics and distortion must be at least 15 dB below the carrier level (-15 dBc). This can be checked during the test by using a spectrum analyzer with a directional coupler on the power amplifier output.

A simple linearity check can be performed at any time by manually reducing the signal source level by 1 or 3 dB and confirming that the output level changes by the same amount. Amplifier linearity is normally quoted in terms of power output at the 1-dB gain compression point, and the specified performance of harmonic generation (in -dBc) is only obtained up to this level. Above it, the distortion rises rapidly (Figure 6.13).

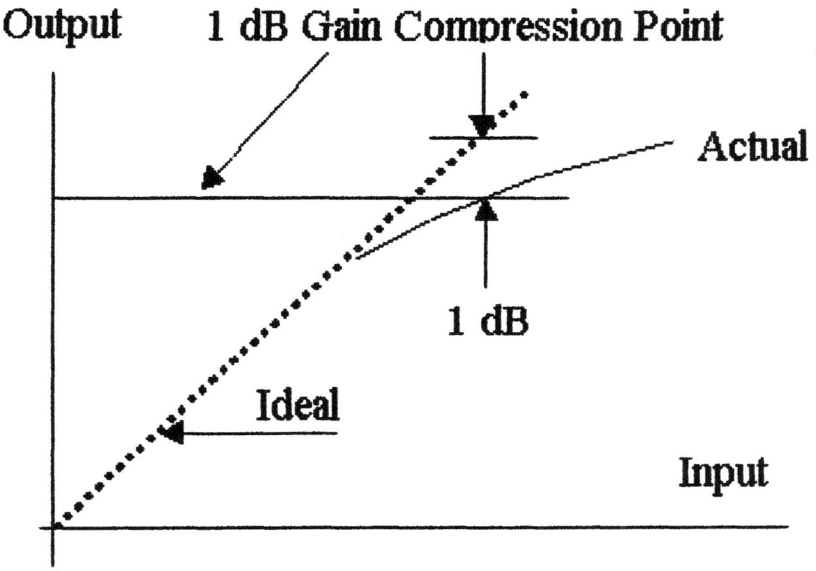

Figure 6.13. **Amplifier Linearity**

Power Gain

Amplifiers are usually specified to deliver their maximum power output for a given input level, typically 0 dBm. The power gain from input to output should be relatively constant over the whole operational frequency range. If it is not, then a higher level of drive signal is needed, typically at the edges of band coverage. This may place an extra demand on the output of the signal generator. The signal generator should have a maximum output level of +10 dBm to compensate for this.

Reliability and Maintainability

Despite the best efforts to improve reliability of equipment, power amplifiers do still fail. When this happens, repair and verification should be accomplished quickly. It is costly to have backup units available, and an amplifier that is out of commission will hold up a large part of the expensive test facility, not to mention product development.

6.4.3 Antennas

Antennas frequently used in EMC work and their frequency ranges are summarized in Table 6.2. For radiated immunity, antennas may be uncalibrated if a field probe is used to measure the generated field. If antenna power is used to calculate the generated field, or the antenna is used for radiated emissions measurements, then the TAF (transmitting antenna factor) and AF (receiving antenna factor) must be verified periodically by having the antenna calibrated. Linearly polarized antennas are preferred to reduce measurement uncertainty. Many times, two orthogonal antenna positions are used in successive test segments, since the polarization for maximum sensitivity to radiated fields may be unknown.

Table 6.2. **Representative EMC Antenna Types**

Frequency Range	Receive Antenna	Transmit Antenna
50 Hz - 200 Hz	Loop	Helmholtz Coil
30 Hz - 50 MHz	Rod, Loop	Wire E-Field Generator
30 MHz - 320 MHz	Biconical	Biconical
100 MHz - 5 GHz	LPDA	LPDA
200 MHz - 40 GHz	Horn	Horn

The receive antenna factor is used to calculate the field strength of a signal impinging on an antenna. By measuring the antenna terminal voltage and applying the antenna factor, the field strength can be calculated. The receive antenna factor (AF) is defined as:

$$AF = E - V - A$$

Where

AF = Antenna factor in dB/meter
E = Incident field strength in dBuV/meter
V = Voltage at input to EMI receiver in dBuV
A = Cable loss in dB

CISPR 16 mandates a VSWR of less than 2:1 on receive antenna systems. This is frequently accomplished by placing an attenuator at the antenna output. Attenuators may range from 3 to 10 dB, depending on antenna VSWR. CISPR 16 further requires that linearly polarized antennas be used. Ferrite chokes are sometimes used on the RE antenna coaxial cable to prevent the cable from electrically unbalancing the antenna.

Transmit antenna factor (TAF) is based on the equation:

$Eo = (30PnGa)^{1/2}/d$
Where

Eo = Field strength in volts per meter
Pn = Net power to the antenna in watts
Ga = numeric antenna gain ($10^{dBi/10}$)
d = distance in meters

Transmit antenna factor is used to calculate the field strength radiated by an antenna at a certain distance. By knowing the transmit power to the antenna and the transmitting antenna factor, the field strength can be calculated.

TAF for free space conditions is defined as
$TAF = 20*\log(f) – AF – 20 \log (d) - 32.0$
Where

TAF = Transmitting antenna factor in dB/meter
f = Frequency in MHz
AF = Receive antenna factor in dB/meter
d = distance in meters

or, for a given distance,
TAF = E -Vm
Where

E = Field Strength in dBV/m
Vm = Antenna voltage in dBV

Assuming the DUT is in the far field of the antenna, good VSWR will result in making best use of the power amplifier's capabilities.

Directivity of an antenna relates to its "beamwidth", the "viewing angle" of the antenna as shown in Figure 6.14. EMC antennas are frequently designed to provide a beamwidth in the primary plane of sixty degrees between the half power points. High directivity increases the field strength at the DUT, while low directivity maximizes the test area that is generated at a constant field strength. To cover a given DUT size, higher gain antennas must be located further from the DUT than lower gain antennas.

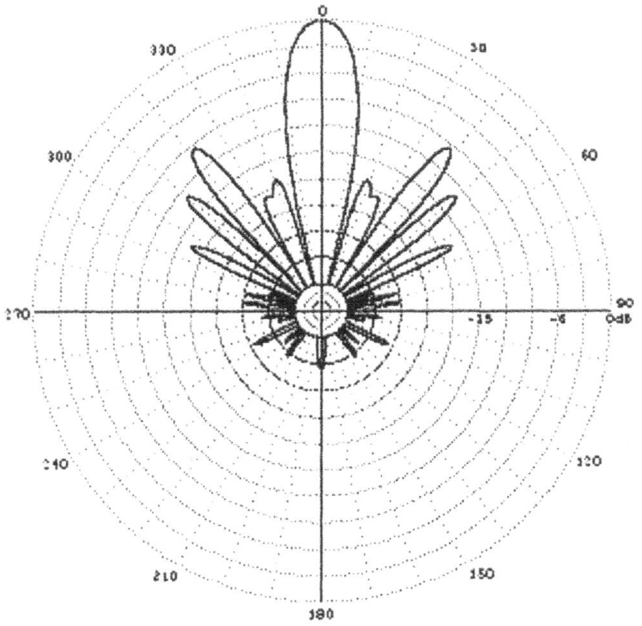

Figure 6.14. **Directional Antenna Pattern**

Antenna VSWR (Voltage Standing Wave Ratio) affects the uncertainty of RE measurements as well as the power required to perform RI tests. The mismatch uncertainty for RE measurements is:

$$U(dB) = 20*\log_{10} (1 \pm |Ra||Rr|)$$

where

U = Uncertainty in dB
Ra = Antenna reflection coefficient
Rr = Receiver reflection coefficient

With the 2:1 VSWR required by CISPR 16, the uncertainty is +0.9 / -1.0 dB.

6.4.4 Field measurement probes

Field measurement probes (Figure 6.15) must be calibrated if they are used to determine the field strength, rather than just monitoring the field strength.

Most probes will respond to the E Field component of an electromagnetic wave.

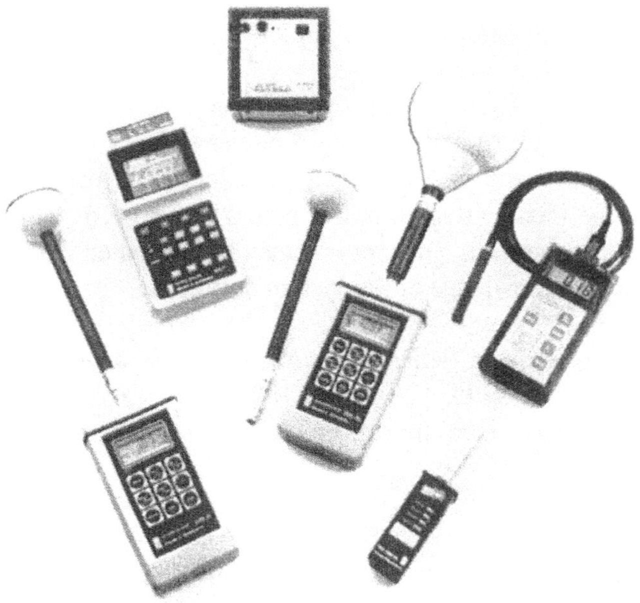

Figure 6.15. **Field Measurement Probes**

To avoid making multiple polarization measurements, probes should be polarization insensitive (commonly referred to as "isotropic") within ± 0.5 dB. Probes should be bus controlled and read. Especially important is frequency range, high dynamic range or remote range switching, to avoid unnecessary test operator intervention. Response to modulated waveforms will avoid the need to establish a field level without modulation prior to applying the modulation. An optical interface will minimize perturbations of the field being measured by the interface. Battery life and charging time are important unless operation from the AC mains is planned. Accuracy should be ± 1.0 dB or better, and settling time (affects test time) of 0.5 μS (micro second) or less.

6.4.5 Power Measurement

When using the calculated field method, power meters and sensors must be calibrated, since they become the amplitude reference for immunity measurements.

Thermocouple-type sensors

Thermocouples operate because dissimilar metals generate a voltage due to temperature differences at a hot and cold (ambient) junction of two metals. Since thermocouple sensors absorb the RF/microwave signal and heat the "hot"junction element, they give the correct average power for all types of signal formats from continuous wave (CW), to pulsed, to complex digital modulation, regardless of the harmonic content, wave shape or distortion of the signal. Thermocouple power sensors were the preferred sensor type for systems with complex modulation formats, as the sensor responded to the total aggregate power across its entire dynamic range. A simulated radar signal peak pulse power may be computed from the average power value and a knowledge of the system duty cycle.

Thermocouple sensors typically have a dynamic range of only 50 dB, from –30 dBm (1 μW) to +20 dBm (100 mW). This restricted sensor dynamic range requires more time to measure the lower power levels.

Diode sensors

Diodes convert high frequency energy to DC by means of their rectification properties, which results from their non-linear current-voltage characteristic. A typical diode detection curve starts near a noise level of -70 dBm and extends up to +20 dBm. In the lower, "square-law "region the diode 's detected output voltage is linearly proportional to the input power (V_{out} proportional to V^2_{in}).

Above –20 dBm, the diode's transfer characteristic transitions toward a linear detection function (Vout proportional to Vin), and the square-law relationship is no longer valid. Traditionally, diode power sensors have been specified to measure power over the –70 to –20 dBm range, making them the preferred sensor type for applications that require high sensitivity. In applications that require fast measurement speed, diode sensors are preferred over thermocouple types because of their quicker response to changes of power.

Table 6.3. **Power Sensor Types**

Characteristic	Thermocouple	Diode
Sensitivity	Average	High
Dynamic Range	50 dB	90 dB
Settling Time	Average	Fast

6.4.6 RF Signal Generator

The RF signal generator may be the device that determines test frequency, unless an external counter is used. If the generator alone determines the test frequency, this quantity needs to be calibrated. If the generator amplitude is not used for the field strength reference, the output amplitude may not need to be calibrated, so long as the field probes or power meters are calibrated and relied upon for amplitude determination.

Types of RF signal sources

- swept – sweeps over a range of frequencies, may be continuous or frequency "stepped"
- signal generator -- adds modulation, produces "real world" signal

Generators with internal modulation capability may be preferred over those that require an external modulator. Sine wave modulation is required for EC testing. Pulse modulation may be desirable to simulate radar signals.

- range - range of frequencies covered by the source
- resolution - smallest frequency increment
- stability - frequency change with time or temperature
- accuracy - how accurately can the source frequency be set
- Harmonic content. Some digital signal generators may have higher harmonic content than their analog counterparts. See Figures 6.16 and 6.17.

Table 6.4. **Signal Generator Characteristics**

	Digital	Analog
Positive	Low Line Power Wide Frequency Range Good Stability	Lower Harmonic Content Inexpensive Less Phase Noise
Negative	Higher Harmonic Content More Expensive More Phase Noise	Requires more Line Power Less Stable

Having a single generator that covers the entire test range will minimize RF switching. Resolution, stability, and accuracy are usually adequate, considering the number of significant digits (frequency is usually reported to three significant digits) used in the data report and graph. Phase noise, harmonic content, and residual FM are usually adequate and the RF amplifier between the generator and antenna is usually more critical for these parameters.

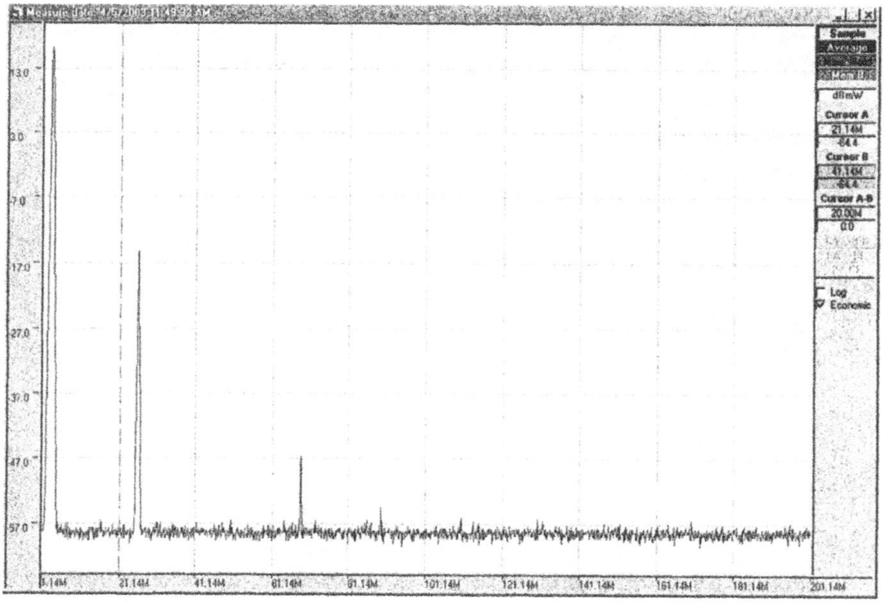

Figure 6.16. **Analog Signal Generator Harmonic Content**

An additional RF generator consideration is transients and overshoot when the internal output attenuator is switched to adjust the output level. Care must be taken when device deviations occur during output level switching, that the deviations are valid and caused by the test waveform, and not caused by switching artifacts.

6.4.7 Electronic Impedance Bridge

Another useful instrument in an EMC laboratory is shown in figure 6.18. This test equipment was originaly designed to be used in the evaluation of antenna systems and has the following features:

- An RF signal generator that covers from LF to VHF, and UHF
- Frequency counter
- An inductance meter
- A capacitance meter
- Measures the SWR as a function of frequency

It turns out that these features make the instrument useful in EMC work due to it's ability to measure parasitic capacitance and inductance, and serve as a source of low power RF energy. Figure 6.19 shows the top panel of the instrument with it's input connections.

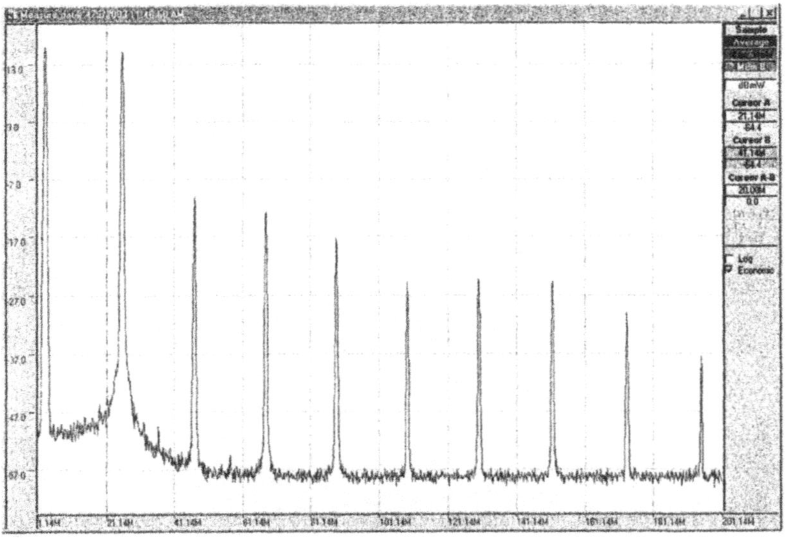

Figure 6.17. **Digital Signal Generator Harmonic Content**

The reader is encouraged to review specification sheets of different types of test equipment that are used in EMC test facilities.

6.4.8 Spectrum Analyzer

An EMI receiver or spectrum analyzer is used for measuring emissions or viewing amplifier harmonics. The EMI receiver features a low noise floor and good frequency selectivity. In some cases noise floor and selectivity are traded for measurement speed and a spectrum analyzer is used instead of a receiver.

Figure 6.18 **Electronic Impedance Bridge**

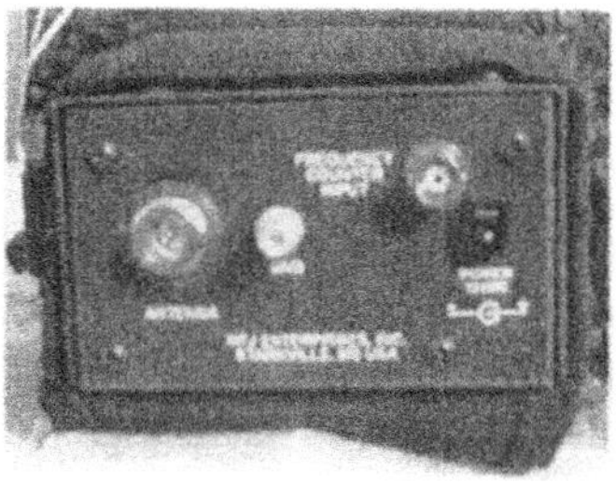

Figure 6.19 **Top Panel of Electronic Impedance Bridge**

Frequency Range	10 KHz to 1 GHz	For typical models (Extended range models can measure much higher in frequency
Frequency Span	Hz to GHz range	User selectable
Resolution Bandwidth	KHz to MHZ range	User selectable
Video Bandwidth	Hz to MHz range	User selectable
Signal Input Range	-100 dBm to 30 dBm	Must insure signal is below the maximum – to avoid potential damage
Demodulation	AM	Enables audio output of AM signal

Figure 6.20 **Spectrum Analyzer Specification Sheet**

The spectrum analyzer screen displays amplitude versus frequency, but can also be made to display amplitude versus time. Amplitude measurements may be scaled linearly for showing signals near the noise floor, or logarithmically to show a large dynamic range of signals. An example of specifications for a spectrum analyzer is shown in Figure 6.20.

Figure 6.21 shows a spectrum analyzer, and Figure 6.22 shows a block diagram. The sweep generator produces a voltage ramp that sweeps the local oscillator and provides a time base for the "x" axis of the display.

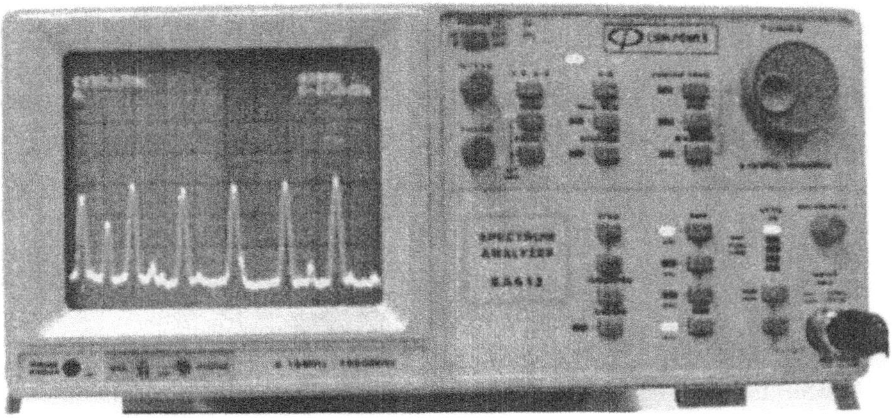

Figure 6.21. **Spectrum Analyzer Displaying several NB Signals**

The local oscillator determines the frequency to which the spectrum analyzer is tuned as it sweeps.

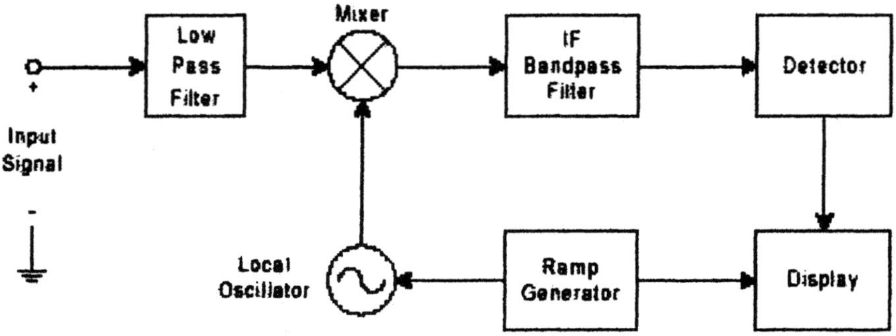

Figure 6.22. **Block Diagram of Spectrum Analyzer**

Note that the frequency selectivity is located after the mixer stage. This means that although signals within the sweep range on screen may appear relatively low in level, higher level signals could be input to the mixer and be outside the frequency range being swept. *This is perhaps the most frequent cause of errors when using spectrum analyzers, and the operator must ensure that the mixer is operating linearly.* The easiest way of verifying this is to add a known amount of attenuation to the analyzer input, for example, 10 dB. All on-screen signals should drop 10 dB in amplitude. If some drop more, we can be sure the mixer is being driven beyond its linear range. Frequency preselection filters, band limiting, or input attenuation are required until we are certain the mixer is operating linearly and not adding distortion products to the screen display.

Assuming the mixer is operating in its linear region, the resolution bandwidth (or "IF" bandwidth) will determine the width of the "window" we are sliding across the spectrum. IF gain will determine dynamic amplitude range, the amplitude points between instrument internal noise floor and the maximum signal that can be input on-screen.

Since mixers have a relatively poor noise figure, sometimes a preamplifier is added ahead of the mixer to measure signals near the noise floor, but caution must be exercised to avoid mixer overload.

Post detector filters (sometimes called video filters) can add smoothing to the detector output. For peak measurements smoothing may be undesireable, and the video filter bandwidth is set between three and ten times the resolution bandwidth. If measurements of coherent noise must be made near the noise floor, sometimes the video bandwidth is set to one-tenth to one-third of the resolution bandwidth.

EMI Receiver/Spectrum Analyzer Detector Functions

Three detector functions are commonly used in measuring emissions, as shown in Figure 6.23. The first is the peak measurement. This measures the maximum value of the emission envelope at each frequency. The second type of detector function is the "quasi-peak" measurement. This yields a time–weighted peak value that is a function of the noise repetition rate. The assumption is that, up to a certain point, higher noise repetition rates are more annoying to the human ear. The last is the average detector, whose output corresponds to the average value of emission itself. The average detector is useful in measuring weak coherent signals near the system noise floor. (Both commercial and government requirements utilize these detector functions.)

Understanding differences among the three detector functions enables one to extract emission signatures from background noise, and can also be used in the diagnostic process. Referring to Figure 6.24, for a specific impulsive noise the peak detector gives the maximum value seen at each measurement "window" across the spectrum. If the same emission is measured using the quasi-peak detector in Figure 6.25, the Quasi-Peak value will be somewhat less than the peak. And if we were to measure the noise with the average value, the average value would be even less than the Quasi-Peak. Knowing these three measurements can help determine which component or system is responsible for the emission.

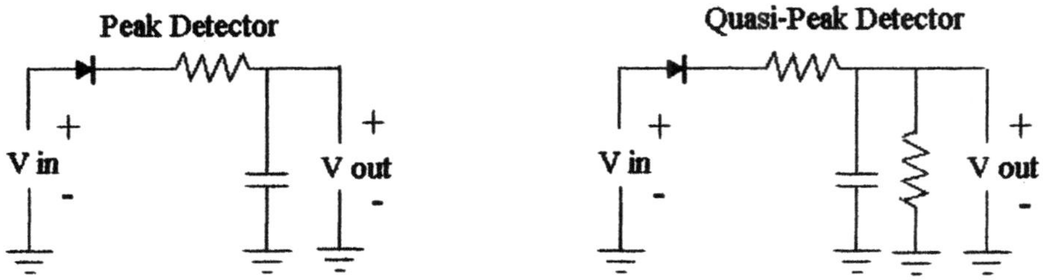

Figure 6.23. **Peak and Quasi-Peak Detector Circuits**

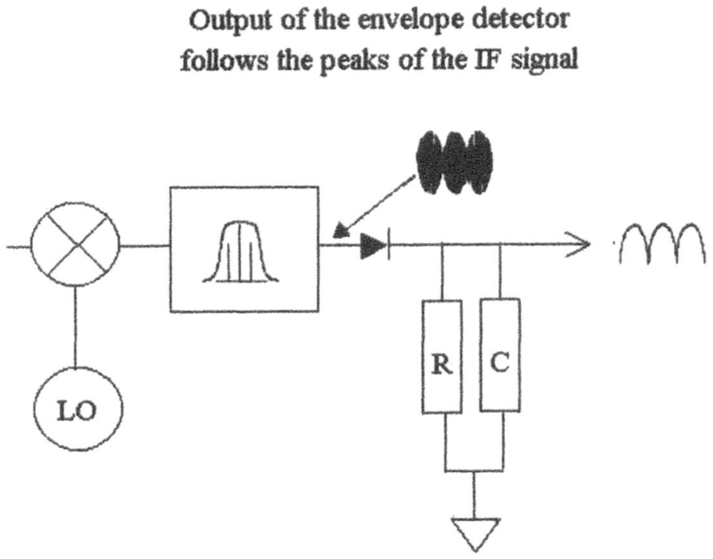

Figure 6.24. **Peak Detector Waveform**

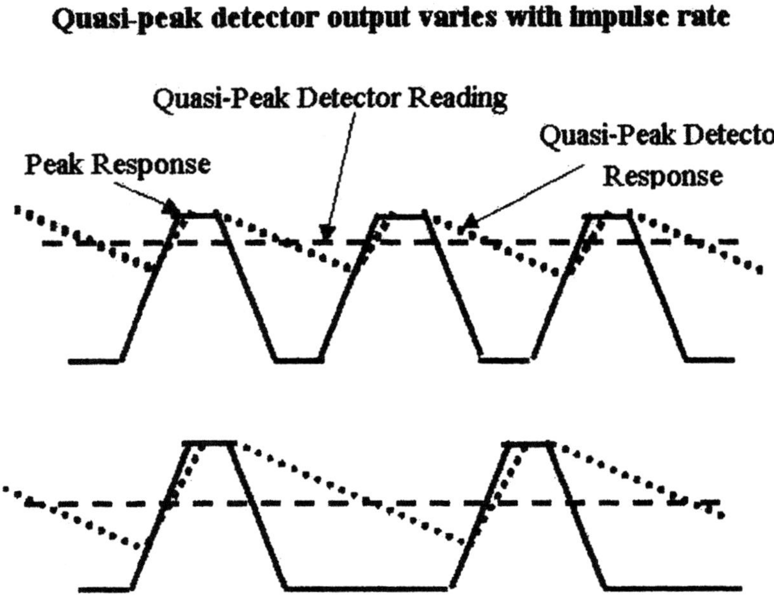

Figure 6.25. **Quasi-Peak Detector Waveform**

6.4.9 Monitoring Equipment

EUT monitoring may employ digital, analog, or communication bus monitors, as well as audio or video links. These are usually connected to the operator's console via fiber optic cable so as not to perturb the field near the EUT. Fiber optic cable may penetrate the shielded enclosure via "waveguide beyond cutoff" pipes, whose diameter is determined by the highest frequency being utilized in the test chamber, and whose length is selected to provide at least as much attenuation as the shielding effectiveness of the shielded enclosure.

Monitoring devices may not themselves deviate under chamber test conditions. In the event of an EUT deviation, monitors that utilize a metallic connection to the wires of the EUT should be then disconnected to verify that they did not cause the deviation. The input impedance of the monitor should be such that it will not suppress an EUT deviation when connected.

6.5 ANALYSIS OF RESULTS

The test report may need to conform to certain standards, for example if the test is intended to demonstrate conformance to FCC requirements, the test report must contain specific information described if 47 CFR Part 2.1033. Standard ETSI EN 301 126 also specifies the format to which the report must conform. One reason for specifying format is that the information may be reviewed by different national regulatory agencies who may not share a common language.

If the report format is unspecified, the test report will at a minimum need to contain the information required by both the product regulation and any applicable quality standards such as ISO9000, etc.

Before beginning a test, an experienced EMC engineer should review the design, parts layout and packaging for best EMC practice.

The test group will need a thorough, considered test plan. This will include the following items.

- Thorough description of EUT operation with schematic diagrams and layout diagrams
- EUT, wiring harness with connectors, and simulator, if required
- Monitoring equipment, fiber optic, video, etc., that will not affect the EUT immunity when attached

- Test frequencies (and levels for immunity) based on those most likely to cause the EUT to deviate from desired behavior, or to interfere with radio services
- Operating modes in which to test the EUT based on analysis of which modes are most likely to fail immunity or emissions guidelines, or have the most severe consequences when they fail
- Production intent software, operating mode or modes most likely to exhibit a deviation or to maximize radiated emissions.
- Information on whom to contact and how to contact them if questions arise

6.6 COAXIAL CABLES

Which system is more efficient?

It is important to understand real transmission line characteristics and how those characteristics affect actual test setup or data from the testing. Let's look at two examples. The first example is shown in Figure 6.26, where we have a 100-watt transmitter operating at 90 MHz, connected to an antenna by 200 feet of RG-58 coaxial cable. In Figure 6.27 we have a 50-watt transmitter operating at 90 MHz, connected to the antenna by a 50-foot section of the same type of cable. Calculations will show how much power is actually being transmitted to the antenna in each instance.

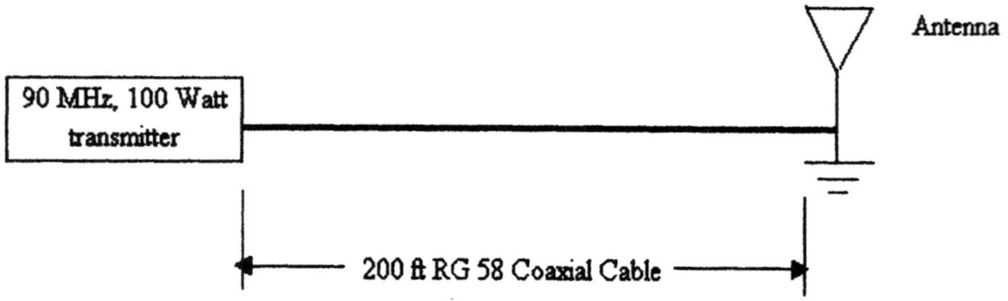

Figure 6.26. **High Transmit Power With High Cable Loss**

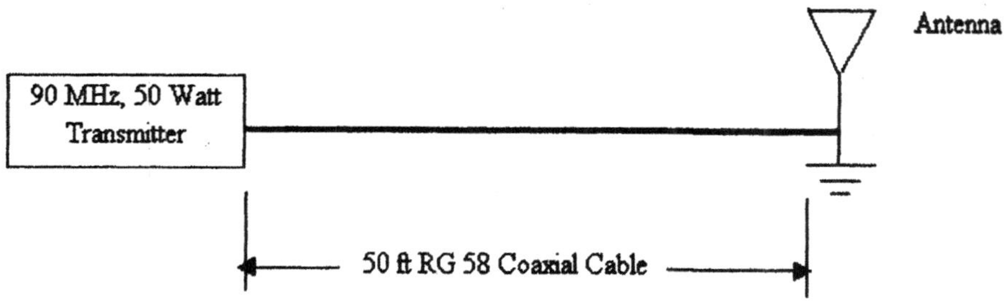

Figure 6.27. **Lower Transmit Power With Less Cable Loss**

In Figure 6.26, the 100 watts travel through the 200 feet of cable before arriving at the antenna. From the tables of common transmission line characteristics in Chapter 4, we see that RG-58 cable should have a loss of approximately 4.5 dB per hundred feet; therefore, the total loss for two hundred feet would be 9 dB. This means that the power delivered to the antenna is 100 watts or 20 dBW (the initial power) minus 9 dB of loss. The calculations for this are also shown below. The calculations show that over 87.5 percent of the power is lost in the cable, and only 12.5 watts reach the antenna.

minus 9 dB = 10 log (power out / 100)
-0.9 equals log (power out / 100)
10 raised to the power of -0.9 = power out divided by 100
0.125 X 100 = 12.5 Watts

In Figure 6.27, the 50 watts travel through the 50 feet of cable before arriving at the antenna. From the tables of common transmission line characteristics in Chapter 4, we see that RG-58 cable should have a loss of approximately 4.5 dB per hundred feet; therefore, the total loss for fifty feet would be 2.25 dB. This means that the power delivered to the antenna is 50 watts or 17 dBW (the initial power) minus 2.25 dB of loss. The calculations for this are shown below. They show that 44 percent of the power is lost in the cable, and 28 watts reach the antenna.

Minus 2.5 dB = 10 times log (power out / 50)
-0.25 equals log (power out / 50)
0.56 equals power out divided by 50
0.56 times 50, or 28 watts

How could the power to the antenna in Case a be increased? Even though the transmitter power is 3 dB greater in Case a than in Case b, the actual power delivered to the antenna is less than one-half the amount in Case a than in b. There are several ways to increase the power delivered to the antenna.

- Use a gain antenna to compensate for the loss in the transmission line.
- Use a larger transmitter or an amplifier to increase the power delivered into the coaxial cable.
- Utilize a transmission line with lower loss at the operating frequency.

Why is it important to understand transmission line loss? The reason is that this information can be used to determine the actual signal level that is being measured by test instrumentation, compared to the level at the measurement antenna. This is shown in the following example:

Assume we have a source of emissions of an unknown amplitude, and we have an antenna and cable connected to the input of a spectrum analyzer that reads 5 dB microvolts at 90 MHz. We also know that there are 400 feet of RG-58 cable connected from the antenna to the spectrum analyzer. It is now a simple matter to determine the magnitude of the level at the measurement antenna.

This can be calculated as follows:

5 dB microvolts + (4 times 4.5) = 23 dB microvolts. If the cable were changed to a lower loss type, the source magnitude would have calculated to a lower level. So we can see that measurement of low level signals requires that cable loss be minimized.

At this point it is useful to review subjects covered in previous chapters that show the importance in EMC testing and analysis.

First is the ability to convert signal levels from linear units to dB. Understanding the relationship between levels expressed in linear units and dB can enhance understanding of problems. It is also important to recognize difference in the formulas to convert volts, power, and current levels to dB.

Second is expressing gain or attenuation as dB. A "positive" dB number is used for gain, and a "negative" dB number is used for attenuation (loss). We have also discussed characteristics of transmission lines and seen tables of information regarding characteristics of those commonly used in EMC laboratories. Any type of EMC data needs to include the complete measuring system parameters, which may include gain from amplifiers, or loss from transmission lines.

This has been a brief introduction to EMC measurement equipment and techniques. It is important to understand that there are government regulations, industry standards, national and international standards, military standards, and contractual requirements. In each instance these may require specific test procedures and data presentation.

6.7 A "VIRTUAL" TOUR OF AN EMC LAB

No understanding of automotive system EMC would be complete without a review and overview of the basic items that would be utilized in an automotive environment EMC facility for both components and vehicles. Vehicle EMC test chambers are shown in Figures 6.28 and 6.29. Much of the same equipment would be found in both vehicle and component facilities.

The following is a list of many common items, and representative figures and/or photographs will be utilized to illustrate the discussions.

One of the most important items in the lab is of course, the antenna used for testing. This is because the antennas are the method to either receive an emitted signal, or to provide an amount of energy to the device under test in immunity testing.

Other sections of the book discussed basic antenna types; Figure 6.30 is a picture of a "biconical" antenna. What can be seen from this is that the antenna looks like two cones connected at their origin point. This antenna is typically referred to as a "bi-con" antenna.

Another key element of any EMC facility is the method to transfer the energy from the antenna(s) to the test equipment. This is accomplished by coaxial cable. Coaxial cable has several aspects that make it the transmission line that is used in a lab facility.

Figure 6.28 **A Vehicle in a State of the Art Anechoic Chamber**
Courtesy ETS-Lindgren, used with permission

In other portions of the text, a detailed discussion of transmission lines was reviewed. Here we will repeat some of the key ones. They are:

1) Flexible usage (can be routed easily)
2) Impedance designed to match most test equipment impedance
3) "Self shielding" aspects – unlike open wire line

The coax types that are most often used are :

- RG-58, a flexible small diameter moderately lossy cable for short runs and jumpers
- RG-59, a small flexible 70 ohm cable for video

- RG-174, a thin flexible cable for mobile radio antennas
- RG-213, a low loss cable with a non-contaminating jacket
- Heliax, a large flexible coaxial cable
- Hard line, a rigid, low loss, high power cable
- Waveguide, a low loss line for microwave frequencies

Figure 6.29 **A Vehicle in a State of the Art Shielded Room**
Courtesy ETS-Lindgren, used with permission

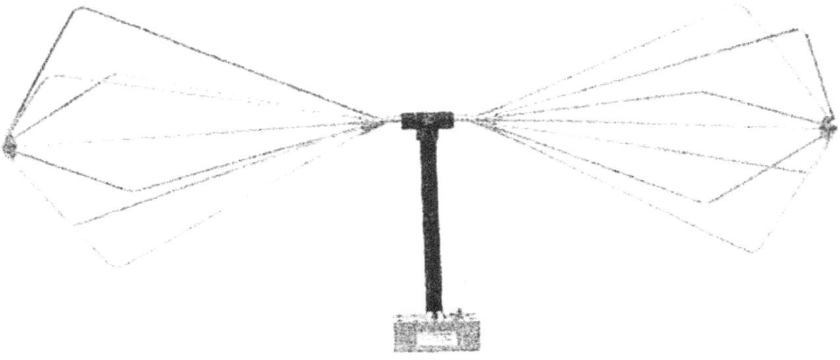

Figure 6.30. **Biconical Antenna**

Examples of some of those are shown in Figure 6.31

The next key item is a fundamental piece of test equipment that anyone who has worked in EMC has needed to study and use, a spectrum analyzer. The details of the operation of a SA have also been discussed in other locations in this text. Figure 6.32 shows a typical SA.

The other item is shown in the following illustration. Referred to in a number of different ways, it is essentially a room that is designed to provide a level of isolation of the external EMC environment from the inside test configuration. This is referred to as a "shielded room". It provides shielding for E fields, and may be used for both RE or RI testing. A typical type of this room is shown in Figure 6.33..

The typical use of an SA would involve setting the analyzer to specific characteristics and then providing an input signal. For most compliance testing requirements, the settings would be specified in the regulations themselves. This is done to ensure that "apples to apples" comparisons occur between the data that is measured and the determination of meeting requirements. Most of the specified settings involve the selection of the resolution bandwidth (RBW) and the video bandwidth. The detailed explanation of each of these has been included in other portions of this text. Other important settings may include the amount of input attenuation that is selected (typically from 0 dB) to 10's of dB. Another setting involves the frequency range that is displayed on the screen, which would indicate the portion of the spectrum that is being displayed.

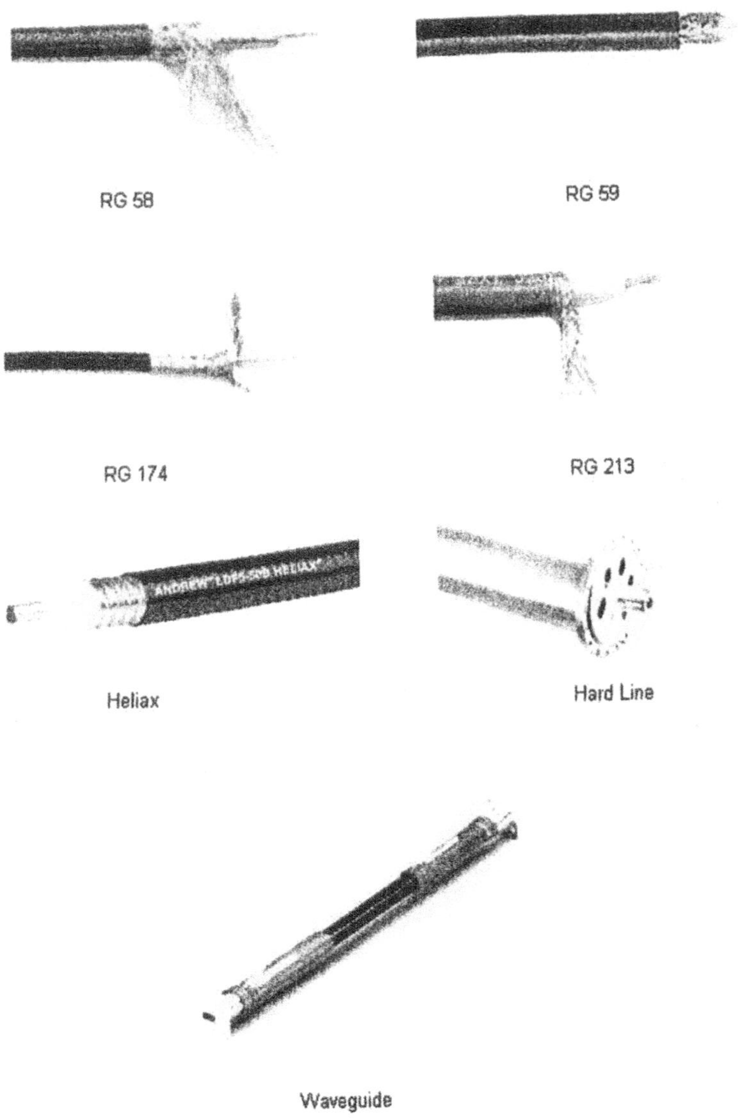

Figure 6.31. **Types of Coaxial Cable**

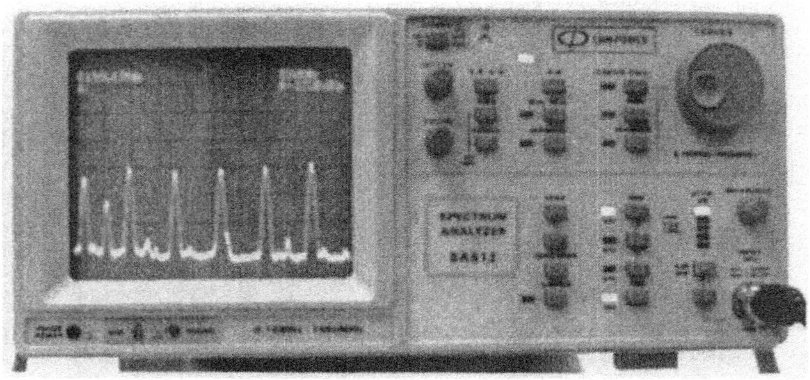

Figure 6.32. **Spectrum Analyzer**

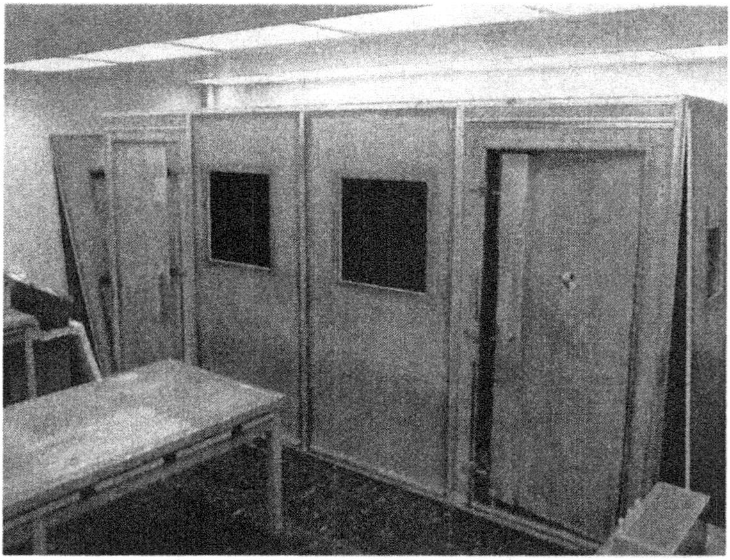

Figure 6.33 **Shielded Room**

Most SA displays would have a range from a few 100 kHz to many 100s of megahertz. The range would be set depending on the specific data desired, and the level of detail required.

Typical screen displays for an SA that is centered on the FM broadcast band may look like Figure 6.34.:

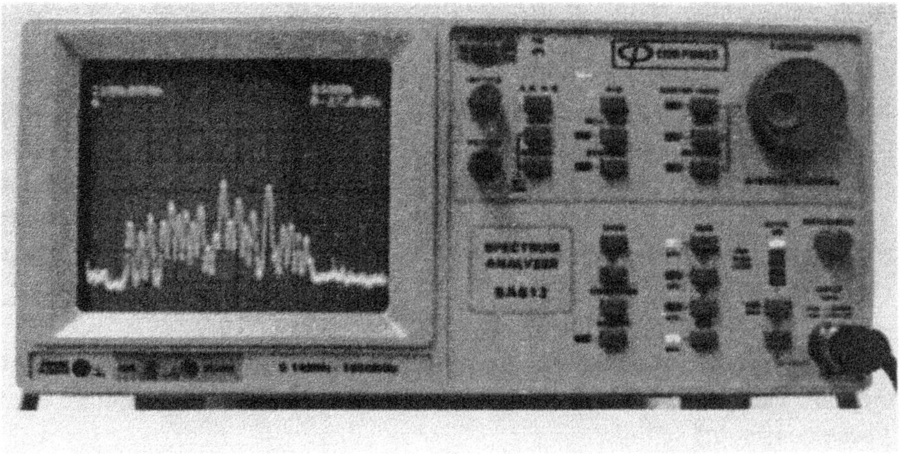

Figure 6.34. **Spectrum Analyzer Shown Measuring FM Band Signals**

This is sometimes a convenient way to display the signals, as it supplies data across a specific band of concern. For this display, we can see that there are many signals, and they vary in their magnitude to each other. This may be due to differences in transmitted power and/or transmitter location from the measuring antenna.

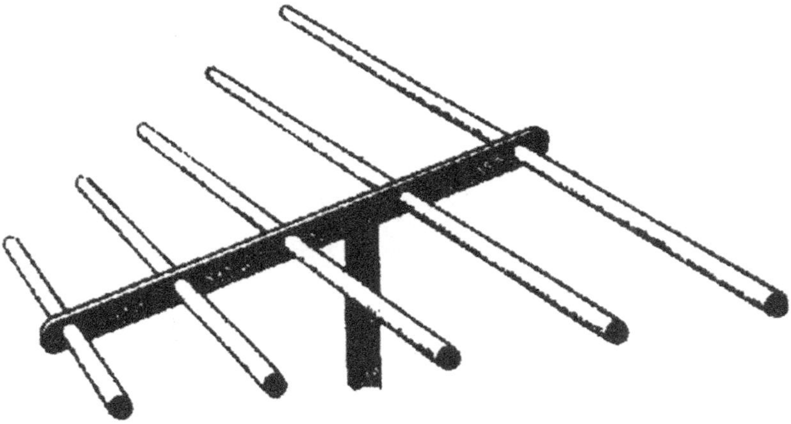

Figure 6.35. **LPDA Antenna (Log Periodic Dipole Array)**

Another common test set up element would be a "gain" antenna, shown in Figure 6.35. These antennas have several characteristics that again make their use important in an EMC test set-up. As discussed in other portions of the text, gain antennas have their primary sensitivity in a particular direction. This means that they effectively can increase the level of the received signal, or can effectively increase the amount of power being transmitted from a transmitter. These may be used to assist in diagnostics even at a vehicle-level RE test. This method may involve focussing the antenna to different portions of the vehicle, and identifying the amount of emissions that occur. Gain antennas are also used to provide additional field levels on systems that are being subjected to immunity testing. This may minimize the requirement for much higher power amplifiers than would be needed if the gain antenna were not used.

To conclude our discussion on common items in an EMC lab and how they are used, we will look at the rest of the typical antennas that are used. These consist of:

Loop Antenna

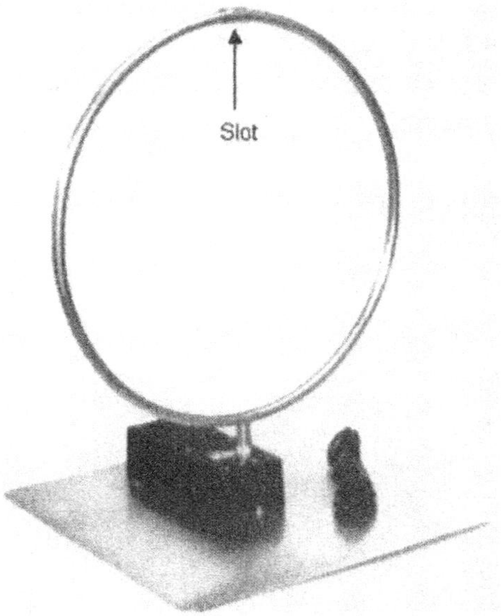

Figure 6.36. **Loop Antenna**

The loop antenna is useful when measuring low frequency emissions, as it reacts to the magnetic field component instead of the electric field. The loop typically also has a specific feature of an electric field shield that encloses almost the complete antenna, except for a small slot at the top. This shield will minimize much of the noise in the lower frequencies, and enable a truer picture of the emissions. A photo of the loop

The last item is the LF monopole antenna. This is a very simple antenna that is used for E field measurements. The monopole, due to its nature, does typically have an amplifier in the base of the antenna, to improve the antenna factor and eliminate the need to switch in matching networks for narrow frequency ranges. Monopole antennas are used typically to about 20 to 30 MHz and are more sensitive to common mode current sources than to differential mode current sources. This antenna is shown in Figure 6.37.

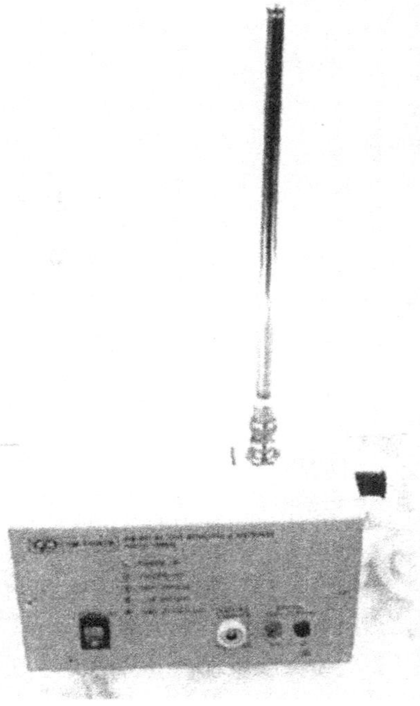

Figure 6.37. **Monopole Antenna**

It is important to know the level of the CE that are being emitted; measuring these levels requires isolation between the device under test and the power lines that supply the primary power to the device. This isolation is accomplished by a device called a "Line Impedance Stabilization Network," abbreviated as LISN. The name for this is pronounced essential "lizzen". LISNs are available in many different values of capabilities with respect to their power and current carrying capability. A typical LISN is shown in Figure 6.38.

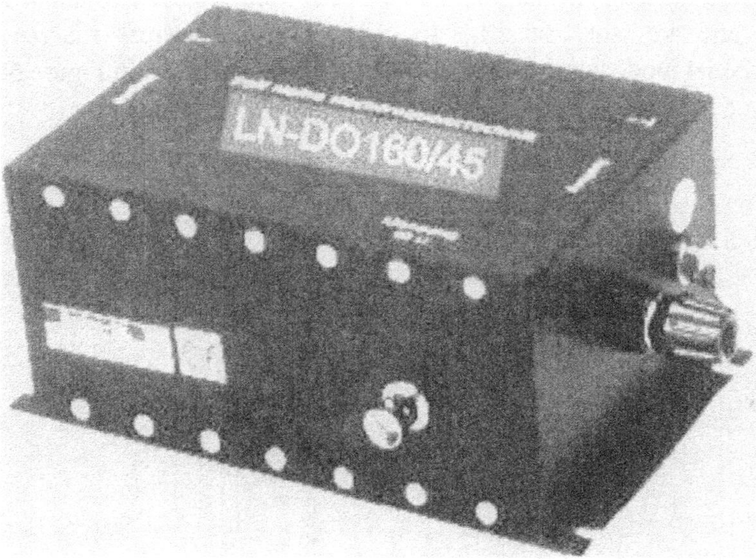

Figure 6.38. **LISN (Line Impedance Stabilization Network)**

The purpose of a LISN is the following:

1. Provide a constant impedance to the DUT
2. Isolate the noise that is put on the power lines from the DUT from feeding back to the power supply
3. Provide a low impedance path for the noise to be measured at the output port of the LISN.

The schematic diagram of a LISN is shown in Figure 6.39:

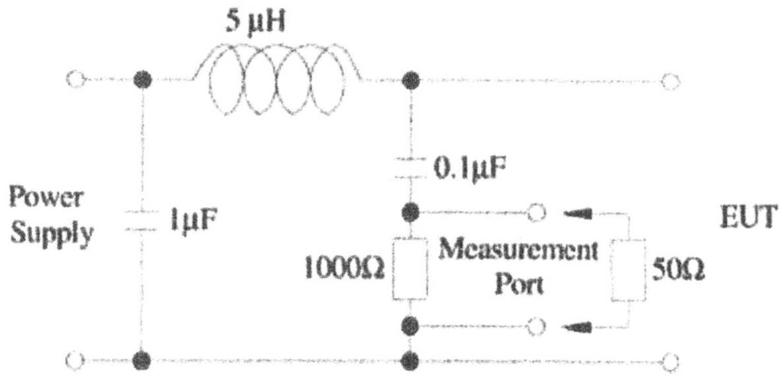

Figure 6.39. **Schematic Diagram of LISN**

The operation of a LISN can be shown from the figure. On the right side is the connection to the DUT, and any noise (which is typically RF content) is prevented from flowing on the power line connections by the 5 uH inductor. The noise is coupled by the 0.1 uF capacitor to the measurement port. On the power supply side of the LISN, any noise that may exist is prevented from flowing to the measurement port by the 5 uH inductor.

The function of a LISN is to isolate the DM current and the CM current from the power supply, and to minimize the impact of the CM current by returning it to its sources. Basically, a LISN consists of a filter network – and almost all devices need some type of power line filtering to minimize CE.

The reader is encouraged to review the basic types of passive filters that can be used and constructed. A brief overview of those filters is included in the following figures.

There are 4 basic types of filters, and their operation can be described in their names. The types are low pass, high pass, band pass, and band reject.

A low pass filter would be constructed to have very little attenuation of low frequency signals up to some intended point. A chart of the operation of a low pass filter is shown below.

A high pass filter would be constructed to have very little attenuation of high frequency signals above a specific frequency.

A band pass filter is a special type that would provide little attenuation between a range of frequencies, and a band reject filter would provide little attenuation above and below a set of frequencies, yet have high attenuation between those points.

These basic types of filters can be constructed in the following ways, shown in Figure 6.40.

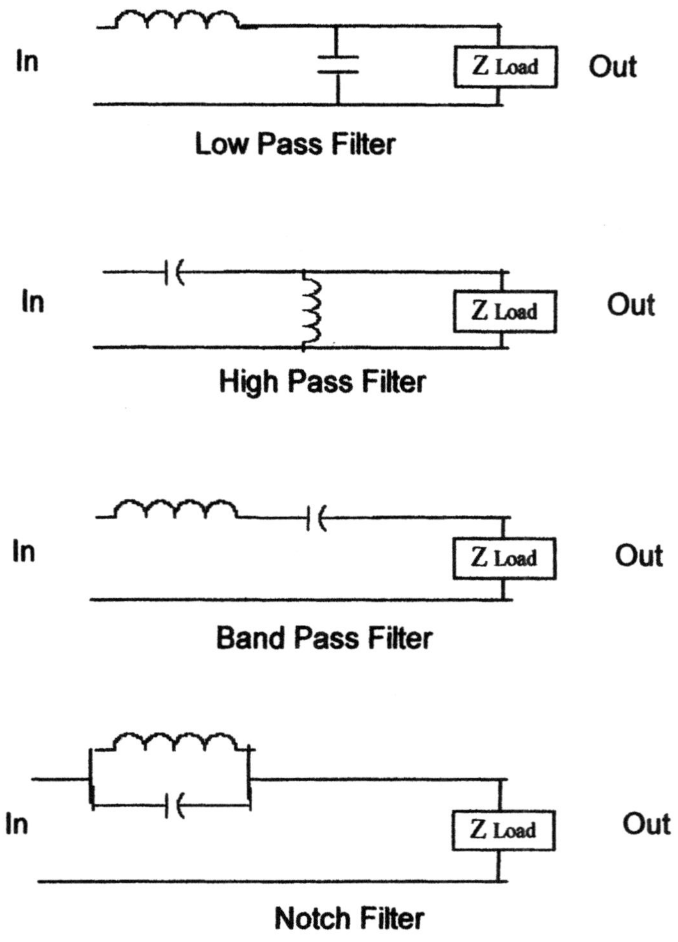

Figure 6.40. **Four Types of Filters**

Chapter 7

EMC Modeling

7.1 THE VALUE OF EMC MODELING

Performing EMC tests is a time-consuming, resource-demanding process. Unfortunately, these tests are often conducted late in a developmental stage, when correcting an EMC problem can be even more difficult and expensive. Validated analytical and numerical methods have the potential to become increasingly important as a process to determine effects of external fields on a car's electronic systems, or anticipating how emissions will develop.

Proper electromagnetic compatibility modeling can reduce development time and the accuracy of the modeling results can be sufficient for planning the layout of cables and components. If the analysis is made during early stages of vehicle design, when layout is more flexible, changes to improve immunity can be adopted without appreciably increasing costs. Early simulation and testing can significantly reduce the time spent testing the final product in the chamber.

Resonances inside the car's bodywork enhance incident fields, which can even rise to several times the external value in some vehicle locations. Therefore, the electric system components (wiring harnesses and electric and electronic devices) must be designed to be immune to such disturbances. Knowledge of these locations can be used to locate cables and electronic systems in regions characterized by reduced field intensities. Conversely, knowledge of vehicle locations that couple well to the outside environment can be used to advantage for locating devices like RKE modules with integral antennas that are designed to couple to devices outside the vehicle.

Modeling can be used to study the coupling of both internal and external sources. The effects of in vehicle portable radio transmitters and both intentional and unintentional receivers (i.e., cellular phones, CD players, GPS, "handi-talkies", etc.) can be as significant as sources outside the vehicle.

The challenge is that EMC modeling is a difficult task due to the nature of EMC itself. Since EMC is the study of "things not on the schematic", much of the complexity in EMC modeling involves determining the relevant parameters to model. Unlike circuit analysis, which is more defined, modeling for EMC is a discipline that is still being designed and developed.

Modeling that has been done to date has primarily involved printed circuit boards and component level devices. These devices are more concise and easier to define than the large systems (such as completing a model for an entire automotive vehicle or system). A search of the references and literature available shows that the techniques have been used with some success on many problems such as crosstalk on a computer board or within a digital device.

The goal of Automotive System EMC modeling is to enable efficient and effective analysis to augment or to completely replace time consuming and expensive testing. This is where modeling has the highest likelihood of impacting auto system EMC work and where the most significant benefits are expected.

7.2 EMISSIONS MODELING

In order to understand the basics of EMC modeling, it is instructive to look at the various areas that have been covered and the specific items in those disciplines. One of the first areas of modeling was to anticipate the levels and types of radiated emissions that could be expected from a device or component.

The reason that this was undertaken first was:

1) There have been numerous studies and data published for years about the occurrence of RE (Radiated Emissions), since it was one of the earliest aspects of EMC.

2) The phenomena responsible for RE are relatively understood. These have been discussed in other portions of this text, and are essentially due to loop areas, common and differential mode current, etc.

3) Many of the actual device and pcb (printed circuit board) layout characteristics that contribute to RE are well documented in the literature.

The task of modeling for immunity is more difficult for a number of key reasons. These include:

1) The requirement to understand component operation and how it may be affected by external sources of energy (which varies from device to device).

2) The coupling path and efficiency of that path in the development of the immunity problem.

3) The manner in which other PCB components modify the interference before it reaches the susceptible component.

Modeling Challenges

Given the need to understand and correctly define the physics of the circuits and the systems, there is still a need to be able to develop the exact numerical tools to define the energy coupling paths by:

1) Understanding the path(s)

2) Quantifying the paths

3) Identifying the applicable corrective actions or countermeasures that can minimize undesired interactions.

From our discussion on printed circuit board layout issues, it should be recognized that the board design and layout are important for the following reasons:

1) The need to understand the characteristics of PCB trace and board interaction and coupling metrics.

2) The need to define and quantify items such as power bus noise that may exist which can contribute to RI problems due to the impact of adding additional energy to a circuit, which already has EMC issues.

3) The requirement to clearly know and define the parameters of the passive elements (lumped and distributed).

4) The need to define various system or component structures that may act as a shield to the external energy. (As we have seen, shielding can substantially change EMC performance characteristics.)

Other contributors to the modeling challenge are the impact of the types of wires and cables that are used, how they are routed, how they are bundled together, and their construction. System or component structures such as frames or cases may actually be acting as energy conductors. This also needs to be considered in the model –this may be a complicated task.

7.3 GOAL OF MODELING

The overall goal of EMC modeling is similar to the modeling work that is done in any other aspect of engineering. The desired approach is that the modeling will assist with key steps in the validation process. One item is the development of the circuit layout. The process of board or circuit layout many times does not include issues that are important to EMC performance. The major contribution of modeling would be to identify potential EMC issues. This is also known as the 80 / 20 rule. The application of this rule to EMC is that 20 percent of the items cause 80 percent of the problems. Modeling should contribute to reducing the work that is expended in trying to identify 100 percent of the issues, of which 80 percent will be determined to not be problems. It is the goal of modeling that the 20 percent of the issues will be able to be clearly defined in terms of their significance to EMC.

Another desired aspect of modeling is the ability to define a model based on a small component, then being able to "upscale" this component to a larger complex system level analysis. This is important to comprehend as it can mean the difference between an accurate system model and results that may be inaccurate.

The intention of EMC modeling should be to take the component level data and use this data to increase the accuracy of the system level analysis.

Some "thought starters" with this point include the following.

1) Use the model to create a "transfer function" from the component level information, then use this transfer function at the system level.

2) Use the model to develop knowledge about system impedance.

3) Be able to model the current flow paths and to quantify the magnitude of the common and/or differential mode currents.

A Possible Process for Modeling

This process can be represented in the diagram information flow paths shown in Figure 7.1..

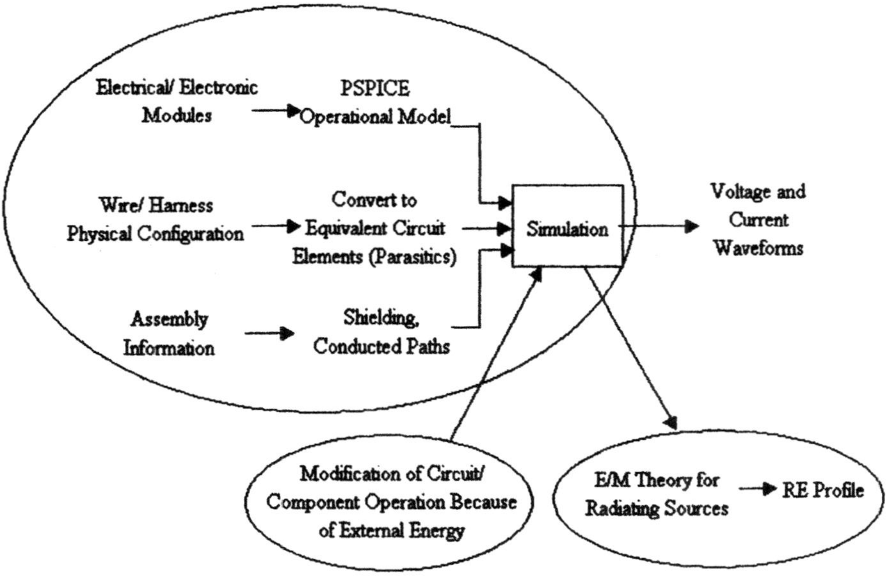

Figure 7.1. **Model Inputs and Outputs**

Vision for modeling

What should be the long approach for EMC modeling? There are some issues that appear to be common between the work and processes used in the studies that take place in the computer engineering work. These include the study of "signal integrity" to understand how power and signals transferred through a circuit.

The process of EMC modeling resembles the following:

1) Use modeling for initial look at system performance.

2) Conduct focussed testing on those areas identified as critical, or do not have good boundary conditions associated with them.

3) Provide feedback on the integrity of the modeling results to continue to make further improvements to the model.

In summary, EMC modeling holds immense promise to eliminate the "design – test – fix – test again" iterations that are common in EMC work. The modeling activities need to emphasize this as a complex process.

Chapter 8

Effects of Cabling and Harnesses

8.1 CONDUCTED EMISSIONS AND IMMUNITY

We have discussed one of the items of an EMC model that consists of the coupling path being radiated through the air or the vacuum of space. This chapter will discuss the fact that is important to remember that emissions and immunity problems can also result from being conducted along some type of wire, cabling, or even conductive portions of a vehicle or system assembly.

8.2 AUTO INDUSTRY EMC APPROACHES

Why are cabling and harnessing important to the automobile industry? The reason is that the industry recognizes that EMC issues can occur on vehicles with electronic modules interconnected by wiring harnesses.

Cabling as interconnections is still very common in the automotive environment. Automotive systems for years to have relied on wiring and harnessing to provide power and signal distribution throughout the vehicle. It is not anticipated that this will change in the near future; therefore, coverage of the effects of all cabling and harnessing is discussed in this chapter. Key elements to understanding harnessing and cabling are the parasitic inductance and capacitance.

8.2.1 Significance of wiring to EMC

Let's look at one of the frequently overlooked, but key items in EMC. In EMC most of the attention is placed upon the obvious components and systems and it is assumed that these are responsible for all the EMC issues. Frequently the connections between the systems and components are not considered, or looked upon as "merely wires", with no particular reason to investigate any deeper into their characteristics. Unfortunately cables and wiring systems can play a large role in EMC issues, and can be the source of many EMC problems! This is because although the wires and cables are "passive devices" on their own, they can create parasitic elements such as capacitors or inductors. These "parasitic" capacitors or inductors that are a result of wires or cabling can cause problems just as if they were intended components included into the circuits. This chapter will discuss the EMC aspects of wiring and cabling, and this relationship to helping minimize EMC issues or unanticipated development of EMC issues.

8.2.2 Role of wiring in EMC

The key elements to understanding how wiring and cabling may contribute to EMC concerns follow. When we connect a component or system to wires, our intention is to bring energy in the form of power or signals into and out of the device or component itself. This connection can also result in the conduction of unanticipated energy or noise into or out of the device. There are two primary types of current that we need to understand for their contribution to EMC:

- Differential mode current
- Common mode current.

Differential mode current is a current that we intend to have, such as the power supply and the power return current flowing on two different wires. This current will flow in opposite directions on each wire. Another example of differential mode current is the signal current and the signal return current flowing along wires line in opposite directions. Sometimes we may have the situation where current flows in several wires or conductors that flow in the same direction along multiple paths. This is called common mode current. The challenge is that both differential and common mode current can cause EMC problems. Each of them has its unique EMC characteristics, including conducting noise along the wire, which acts as an "antenna" to receive external energy or to cause energy to be radiated from the system.

8.2.3 Early vehicles wiring

The impact of automobile wiring harnesses contributing to EMC issues on today's vehicles can be significant. If we compare earlier models of vehicles, Figure 8.1.a, note that there were only a few electrical and no electronic devices (except for the radio – if there was one installed). If we look at the vehicle wiring, we notice that the wires were primarily connected to only switches to control various lights and the ignition system. These vehicles had approximately 150 feet of cable, and the wiring weighed only about 10 pounds. If we now contrast that with a recent model vehicle, Figure 8.1.b, we can see that there is a significant difference in both the number of items on the vehicle and the complexity of the wiring that comprises the harness. Today's vehicles can have well over a mile of cable that may weigh over 100 pounds! The complexity from the EMC standpoint is that, in comparison with the older vehicles, today's vehicles have several characteristics. This is because:

- We did not have many electrical or electronic devices
- The devices that we did have were connected by relatively simple harnesses
- As a result, there were minimal EMC issues due to wiring.

Figure 8.1.a, **Older Vehicle wiring Harness**

8.2.4 Vacuum cleaner incident

Another issue with regard to conducted issues is power supply loading causing primary voltage and/or current to go below its required level due to a component being switched on or off. There are many tests that are run in the automotive environment to determine the impact of positive and negative voltage transients. What is interesting is that this is not unique to automotive systems, as there is a case of a 1985 Spacelab mission run by NASA that illustrated the consequences of this type of system interaction and component performance. During a Spacelab mission, the crew decided to use a certain vacuum cleaner instead of the approved lab vacuum cleaner. This caused a voltage level to drop in the flight control systems and caused a flight computer to turn off inadvertently. The bottom line is that for this application, the vacuum cleaner had not being fully tested and should not have been used.

Figure 8.1.b. **Typical Modern Vehicle Wiring Harness**

8.2.5 Common Mode and Differential Mode Current

A key concept in EMC is understanding when current is flowing and where it is not. If this sounds like a basic statement, the reason is - it is!

There are two types of current that can flow in a wiring harness. At this point it is important for us to examine some of the characteristics of these two types. They are called "common" and "differential" mode currents. Common mode current means:

"Current flows along two or more conductors in the same direction at the same time." This is shown in Figure 8.2.

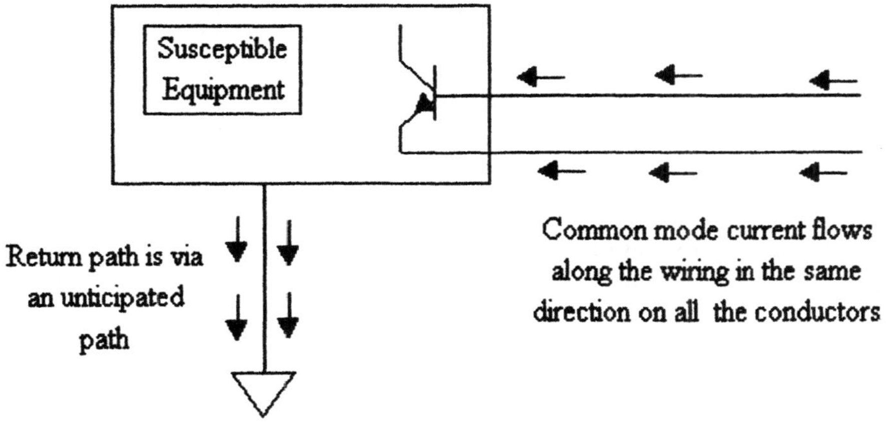

Figure 8.2. **Common Mode Current**

8.2.6 RF emissions and immunity

In "differential mode" current flow, it means that the current flow is 180 degrees out of phase with other current(s).

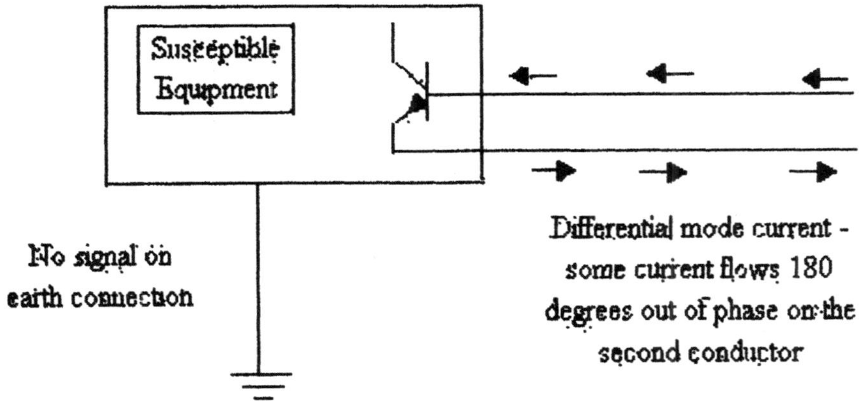

Figure 8.3. **Differential Mode Current**

This is identified as Id, and the common mode current is identified as Ic. In a pure differential mode current flow, Id would be equal and 180 out of phase with each other. In practice, the total common mode current would be the sum of both of the Ic shown on the diagram. This would then result in a total flow on each of the lines of I_1 and I_2. These would be the algebraic sum of Id and Ic, as shown in the equations:

While seeming to be insignificant in themselves, a study of common and differential mode current can assist in diagnosis of EMC problems. This is because each of the currents creates different types of conditions. Ideally, our circuits and systems would have differential mode current, with clearly defined paths. This is not always the case, and we have experience with common mode flow in many instances.

The advantage of differential mode current is that it follows predictable (and the intentional) current path. Common mode current will follow the path of least impedance, even if that path results in unanticipated system or component operation. This can be even a result of current flowing into a "output" that was planned. If this does occur, there make be paths of current that cause incorrect operation.

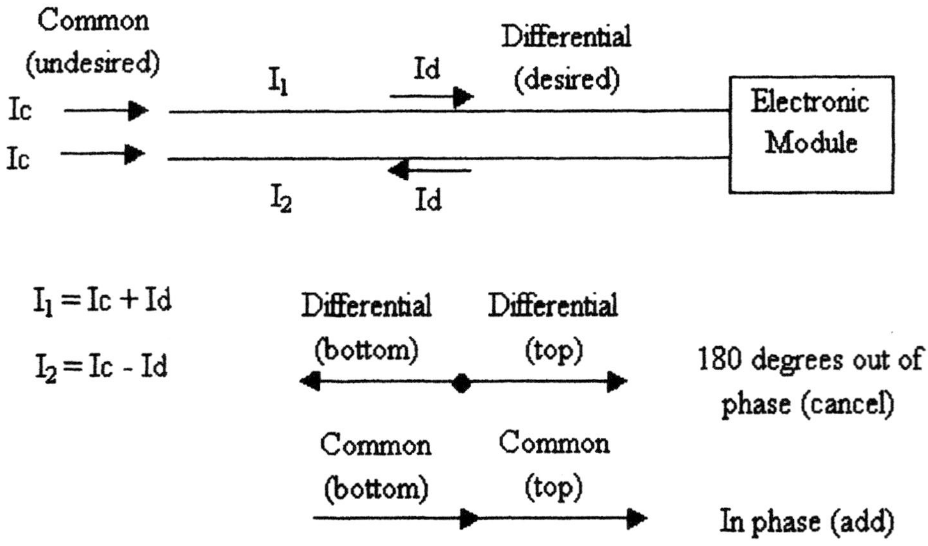

Figure 8.4. **Differential and Common Mode Current**

The following is a representation of the difference between common and differential mode current.

Another aspect of differential mode current is that if the magnitudes are the same and 180 degrees out of phase, the magnetic fields due to each current flow will cancel. The result that if we integrate in the region about the total conductor, the magnitude of the magnetic field will be equal to zero.

Let's look at the radiated emissions that would occur from each type of current flow, and how to minimize their impact.

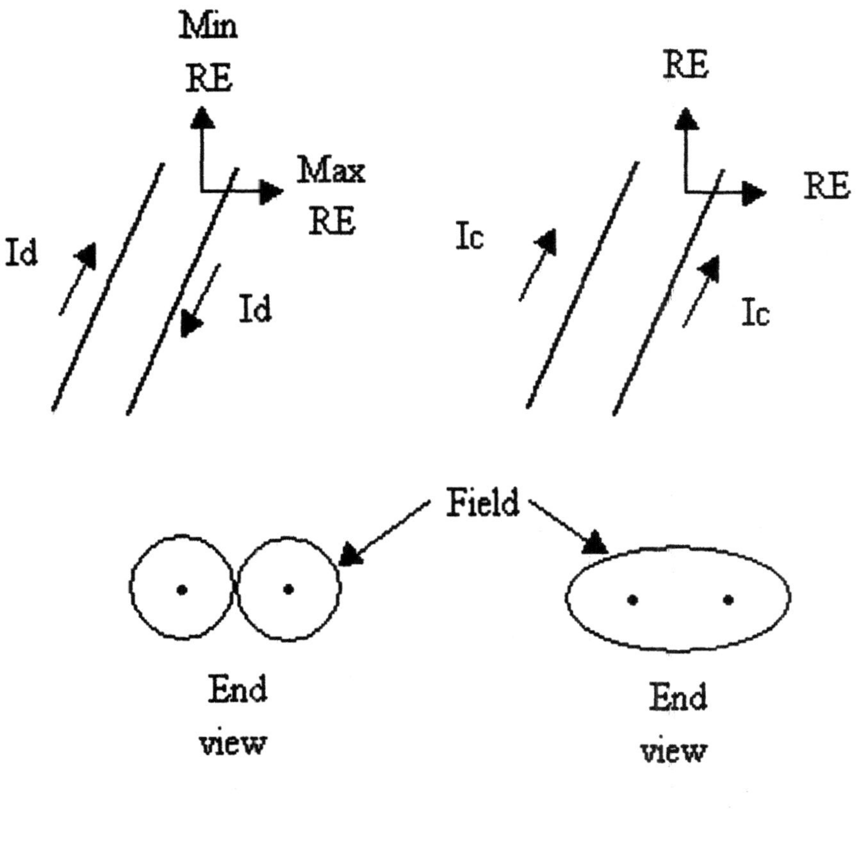

Figure 8.5. **Differential And Common Mode Radiation**

Table 8.1. **Common Mode and Differential Mode RE**

	Correction	Field Intensity
Differential Mode	Reduce Current Reduce Line Length	Not Rotation Sensitive
Common Mode	Reduce Current Reduce Loop Area	Cable Rotation Sensitive

Table 8.1 shows some of the radiated emissions characteristics of both common and differential mode current. If we look at a diagram of these current types, we see the following:

- For the DM current, the minimum RE field occurs at 90 degrees from the plane of the conductors, and the maximum RE field occurs in the plane of the conductors. An end view shows a sideways figure "8" that surrounds the conductors.
- For a loop of less ¼ wavelength for the DM current, the minimum RE field occurs at 90 degrees from the plane of the conductors, and the max RE field occurs in the plane of the conductors. An end view shows a sideways figure "8" that surrounds the conductors. If the loop length is ¼ wavelength, the pattern changes by 90 degrees.
- In the case of the CM current, we have a symmetric RE field that surrounds the conductors. This is due to Ic, and acts similar to the way an antenna operations (CM current is sometimes called antenna current due to its efficiency of RE). The end view for CM current looks like a loop around the conductors.

Let's summarize the characteristics of RE due to RF current.

Table 8.2. **Characteristics of RE Caused by RF Current**

Type	Effects Analyzed By	Caused by
Differential Mode	Transmission Line Models	"Desired" Current
Common Mode	Various Effects	"Undesired" Current

Table 8.2 illustrates that we can have RE caused by both DM and CM currents, and the analysis of the fields from each of those is different. This also illustrates that the CM currents that cause RE can be difficult to analyze, another reason we don't want those types.

8.2.7 Ways to measure RF current

Since it is important to know the flow of noise or RF current within a circuit or along a conductor, this can be accomplished by using various types of probes. These probes typically are clamped or installed around a conductor, and measure the RF current due to the magnetic field that is created around the conductor (right hand rule). The probes can be calibrated to provide either relative or absolute indications of the current. They are installed as shown in Figure 8.6 and would give the results as shown:

From Figure 8.6, it can be seen that that the DM current is surrounded by the probe. Each of the conductors of the DM current will produce a magnetic field, and each magnetic field is the opposite direction from the other. This will result in a current probe measurement of zero, since there is DM current.

In the case of measurement of CM current, shown in Figure 8.7, the measured value will not be zero, since it is based upon the algebraic sum of each of the magnetic fields, which is based upon the current flow.

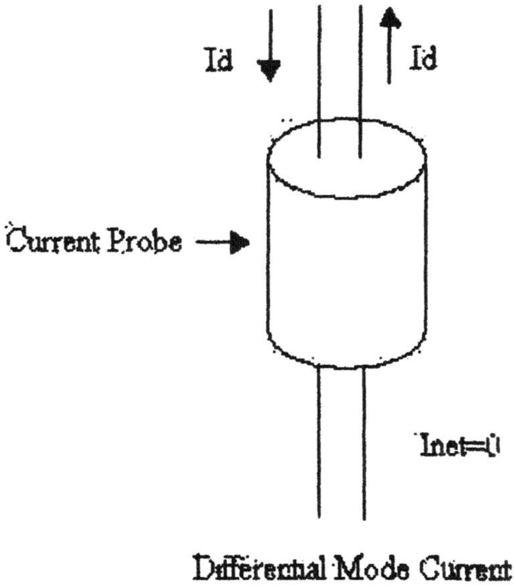

Figure 8.6. **Measuring Differential Mode Current With A Probe**

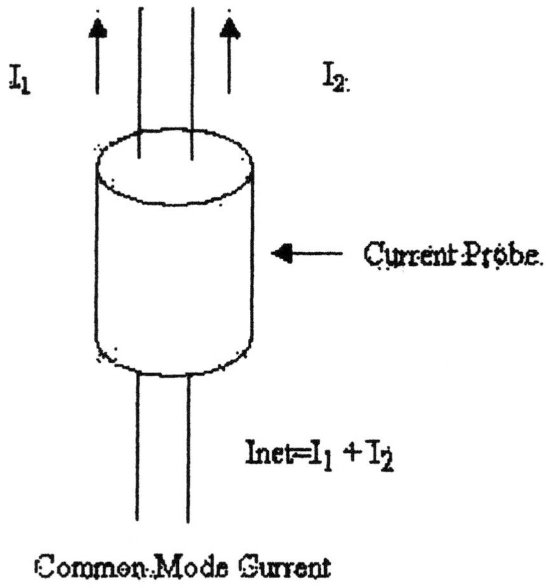

Figure 8.7. **Measuring Common Mode Current With a Probe**

This illustrates that use of current probes can be an important diagnostic tool, provided that it is understood how they operate and what type of measurement is obtained.

Figure 8.8 shows a practical example of the use of a current probe for RF frequencies is in verifying the functionality of a "ground strap". These types of connections are used frequently in an attempt to reduce emissions from a component or system, and many times they are not effective, though it is not always understood why. A quick way to determine how effective the "ground strap" is would be to measure the current flowing through the connection (perhaps focussing in on the specific frequency of noise that is attempted to be minimized). If there is no current flowing, then this indicates that the "ground strap" is not working as intended. (The authors have seen many examples of the need for "ground straps" – and whether the straps were installed or the straps were removed, the system had the same characteristics.) If this basic type of measurement would have been made, the cost, timing, and manufacturing impact of the installation of these connections could have perhaps been averted!

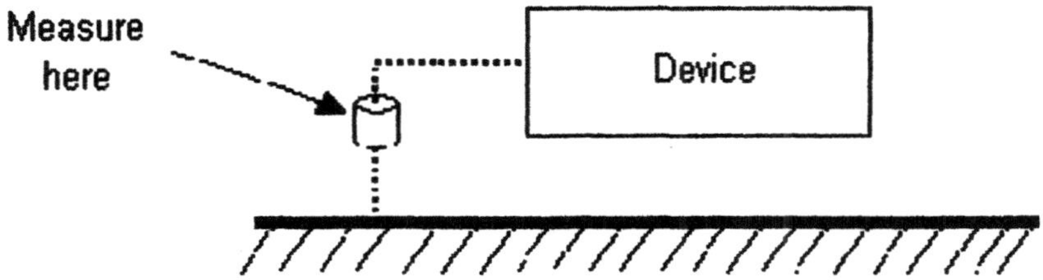

Figure 8.8. **Measuring Current in a Ground Strap**

There are low frequency versions of the clamp on current probe, and these are used in diagnosis of problems with 60 Hz currents. These devices are simple to construct and essentially consist of a coil of wire (which serves to pick up the magnetic field) and a basic meter. There would also be a diode in the circuit that would rectify the AC current, allowing the DC meter to read a magnitude of voltage that is proportional to the current that is being sensed.

The electrical schematic and general physical representation is shown in Figure 8.9.a, and a photograph is shown in Figure 8.9.b

Figures 8.10.a and 8.10.b show the cancellation effect of differential current. In Figure 8.10., if only one line is measured, the actual current flowing one way will be displayed. If both lines are within the current probe as in Figure 8.10.b, the opposing currents will cancel, and no net current flows.

The concept of differential and common mode currents can be carried into higher frequencies that may be causing EMC issues in automotive systems. In this case, the current has been created by an external source of energy, such as a nearby RF transmitter. Several key elements contribute to the magnitude of the possible problem:

- The frequency of the external field.
- The characteristics and the physical dimension of the conductors in the system that is being affected.
- The direction and magnitude of the external field.
- The source and the load impedance in the system that is being affected.

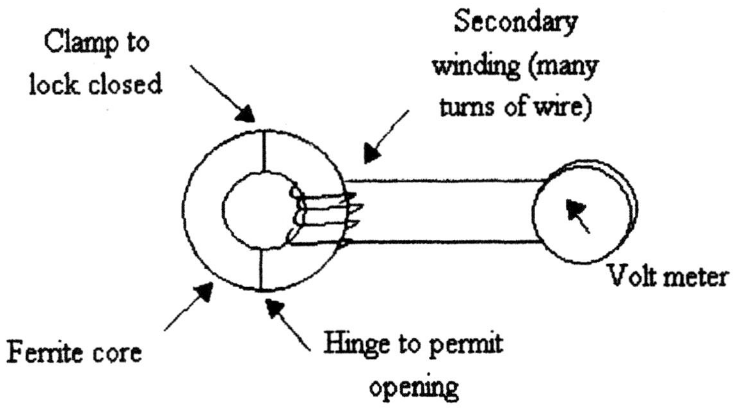

Figure 8.9.a . **Current Probe Electrical Schematic**

Figure 8.9.b **Commercially Available Current Probe**

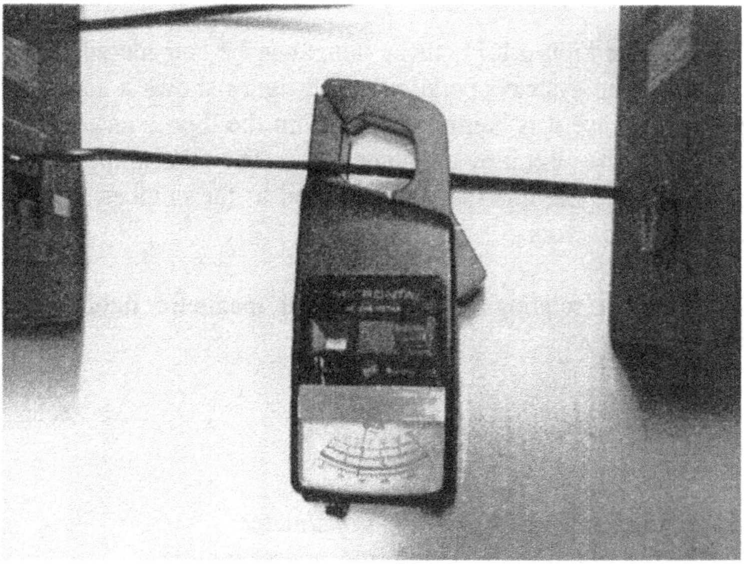

Figure 8.10.a **Measuring the Current in One Wire. Current is Displayed on the Meter**

Figure 8.10.b **Measuring the Current in Both Wires. The Net Current is Zero**

This can be shown as follows:

If we look at Figure 8.11, these items can be considered with respect to their impact upon system operation. The figure shows a source of external energy (in this case it is being radiated from the dipole antenna), and this is separated from the circuit by a distance (D). The other important aspects are the lengths of the harness that is connected to the devices, and the spacing between the conductors.

The equation relating the electric and magnetic field strengths is as follows:

$\mathbf{E} = 377 \times \mathbf{H}$
Where

E is the electric field strength in Volts/meter
H is the magnetic field strength in Amperes/meter
377 Ohms is the impedance of free space

From these equations, G represents the gain (in linear units) of the transmitting antenna, Pt is the total power that is being radiated, and d is the distance from the antenna to the circuit that is being affected.

After a determination of the E field strength has been made, it is an easy conversion to compute the magnitude of the H field strength. This is related to the E field divided by the impedance of free space (assuming that the distance d places the circuit in the far field of the radiated energy.)

In summary, the important items about common and differential mode currents are the following:

Both types can cause radiation from a circuit.

Differential mode in theory has 100 % field cancellation, which makes it less likely be radiated energy, compared to common mode.

Both of these conditions can result in introducing external sources of energy into a circuit or a system. This can be caused by wire length or the creation of loops.

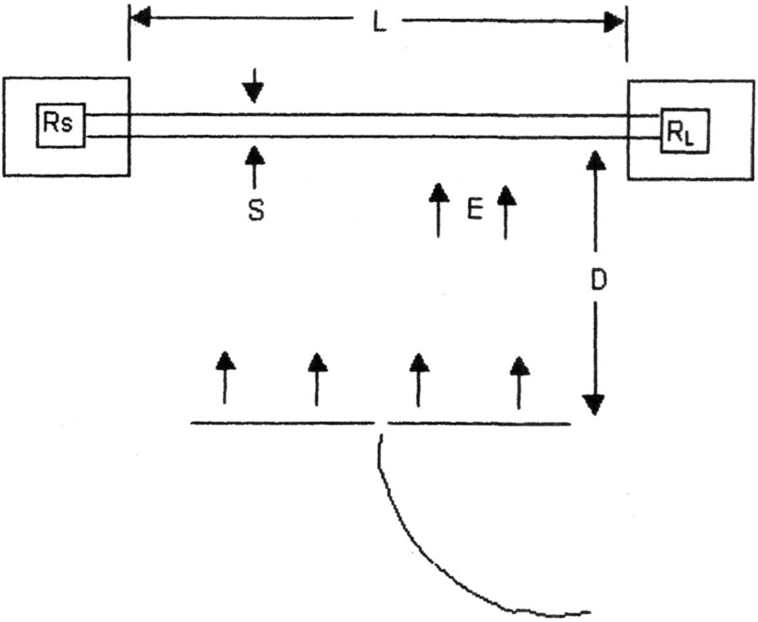

Figure 8.11. **Wiring Harness in the Field of a Dipole Antenna**

8.2.8 Differential mode RE levels

Let's now look at the radiated emissions that can be expected from circuits with DM current. Although there is high degree of cancellation with this type of circuit, there are aspects of the geometry of the circuit that will cause a field to be radiated.

The relationship between the field strength (in the far field) and current levels in DM circuits can be expressed as the following:

$$E = 131.6 \times 10^{-16} (f^{2} A I) (1/r) (\sin \theta)$$

Where:

E is the field strength
F is the frequency of the DM current (the noise current)
A is the area of the loop of the DM current
I is the value of the current
r is the distance from the loop

Sin θ is the angle between the loop and the measurement point (for worst case, make sin θ = 1)

Another important aspect is that the length of the loop of the source needs to be less than ¼ wavelength of the current.

This equation also assumes that the loop is located in an open space area. If the loop is located above a reflecting surface, the value of the constant is doubled to 263 x $(10)^{-16}$, as the energy that would have been transmitted towards the surface will be reflected in the other direction, and would be in phase with the original emission.

Let's look at an example of this type of emission as shown in Figure 8.12.

F = 96 MHz
A = 3.14 x $(10)^{-4}$ meter
I = 10 ma
R = 3 meters (far field = yes)
The diameter of the loop is 20 mm

By the equation, the E field strength is 0.127 mV/m or 42 dBuV/m.

This now tells us the value of the emission from the loop, measured at 3 meters from the loop. If we wanted to reduce the magnitude of the emission, what could be done? Options include reducing the area of the loop. In an automotive application, this may be possible by changing some of the routing of the wire harness, or re-orienting some of the wires in the harness. Another item that can be done is to reduce the current in the circuit. This will also reduce the field strength. The last item, which from an E field standpoint would be the best, is to be able to position the device or circuit such that sin theta is equal to zero. This probably the least practical, however.

From a realistic standpoint, it can be difficult to have only DM current since most of the time the small system imbalances produce CM current, which is then difficult to diagnose.

8.2.9 DM related to design of circuit

We have discussed earlier that it is desireable to have DM current whenever possible, as CM current can have more impact upon EMC RE issues. We can also describe the impact of this CM current and field strength by an examination of the geometry of the circuit.

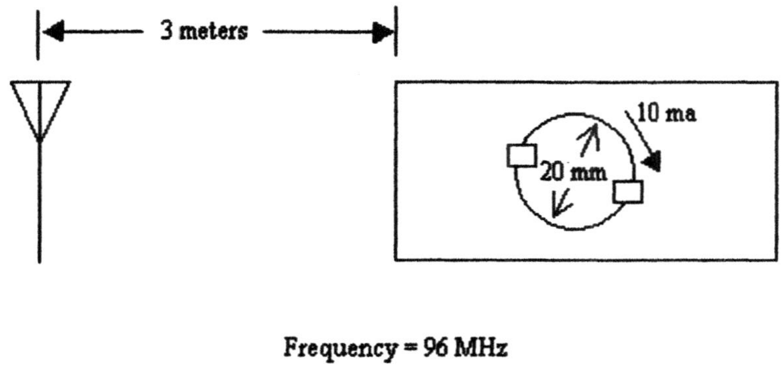

Figure 8.12. **Differential Mode Radiation**

The problem with CM radiated is that this typically results from the interconnection of circuits with different noise voltage levels. This results in CM current flow on the cables/wiring. Figure 8.13 shows this condition:

In this figure, the CM current flows along multiple conductors in the wiring configuration. Essentially, this looks like the system shown in Figure 8.14.

This means that the radiation is from the wire interconnect between the two devices. If we now simplify by one more step, we can see in Figure 8.15 that what we have created is a vertical antenna, with the antenna current now represented by the CM current.

The equation for this relationship is as follows. The e field strength is due to the following conditions, that of the value of the current, the frequency, and in this case, just the length of the conductor can have a major impact.

$$E = 4\text{-}\pi \times (10)^{-7} (f\, I\, L)(1/R) \sin\theta$$

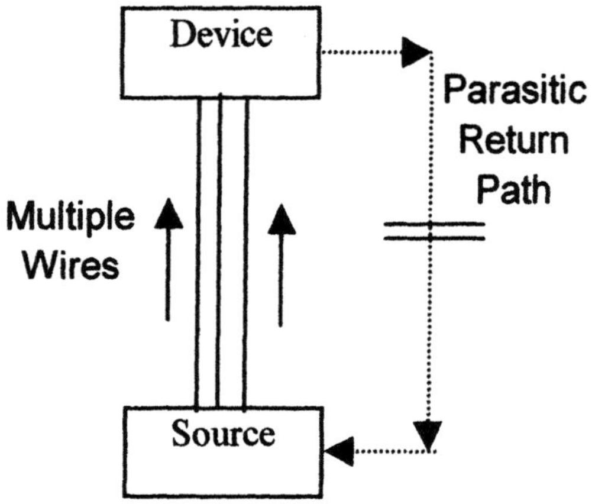

Figure 8.13. **Source and Device With Common Mode Current**

The parameters are the same as before, except that L is the length of the conductor.

Using the parameter that we used before, except for this geometry, yields an E field of 8.04 mV/m or 78 dBuV, which is 36 dB greater than before! This is only due to the change from DM to CM current. So as we can see, circuit type can have a major impact.

8.2.10 Cable Shielding

The major part of the coupling path for interference from the product to the environment and vice versa is through its connecting cables. These form efficient transducers with the outside world, via conduction at low frequencies and via radiation, particularly around their resonant frequencies (at which the cable length is a multiple of a quarter wavelength). These cables may actually intentionally carry high-frequency signals, such as data or video, but a more potent interference source is common-mode noise coupled onto the cable at the interface, and flowing in all its conductors or in its screen, which may not be directly related to the signal. A major part of EMC design is therefore concerned with the interfaces between the unit and its cables.

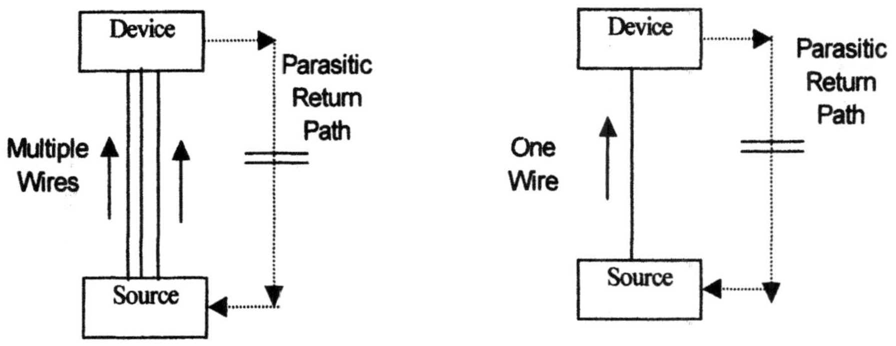

Figure 8.14. **Equivalent Circuit**

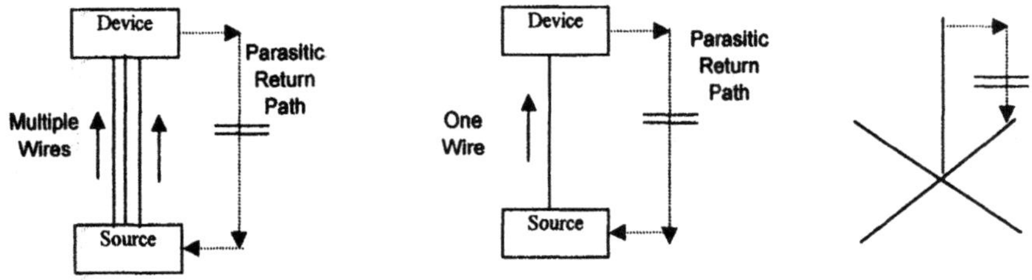

Figure 8.15. **Vertical Antenna Equivalent**

One primary source of RE is unshielded or improperly shielded cables. There are four common types of shielding:

1. braid
2. flexible conduit
3. rigid conduit
4. spirally wound sheets of high permeability material

Of these four, braid is relatively lightweight and easiest to handle. It is important to note that the shielding effectiveness of a cable shield depends on the characteristics of the shield material and the manner in which the shields are terminated.

When terminating a shield, it is important that the termination provide a low impedance path for noise currents. Shield terminations fall into two categories: pigtail termination and 360 degree shield termination (sometimes referred to as RF backshell termination). The 360 degree shield termination provides a low impedance path and preserves shielding integrity of the enclosure or connector to which the shield is terminated. This type shield termination is much preferred. A pigtail termination is the least preferred method of shield termination because, at RF frequencies, the inductance of the pigtail becomes such that the shielding effectiveness of the cable shield is negated. If, however, pig-tail termination is unavoidable, keep the pigtail as short as possible. Figure 8.16 shows examples of pigtail termination and RF backshell termination.

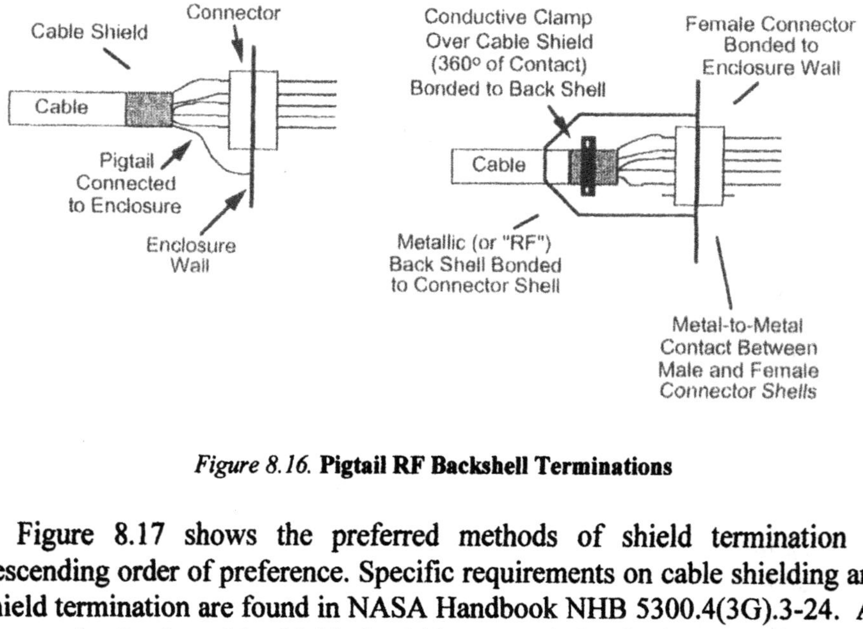

Figure 8.16. **Pigtail RF Backshell Terminations**

Figure 8.17 shows the preferred methods of shield termination in descending order of preference. Specific requirements on cable shielding and shield termination are found in NASA Handbook NHB 5300.4(3G).3-24. As a general rule, the cable shield should be grounded at both ends. Also, cable shields should never intentionally carry current. The exception to this rule is coax cable, in which the outer shield serves as the return conductor. Coax should be used only for signals where the lowest signal component is above approximately 100 kHz.

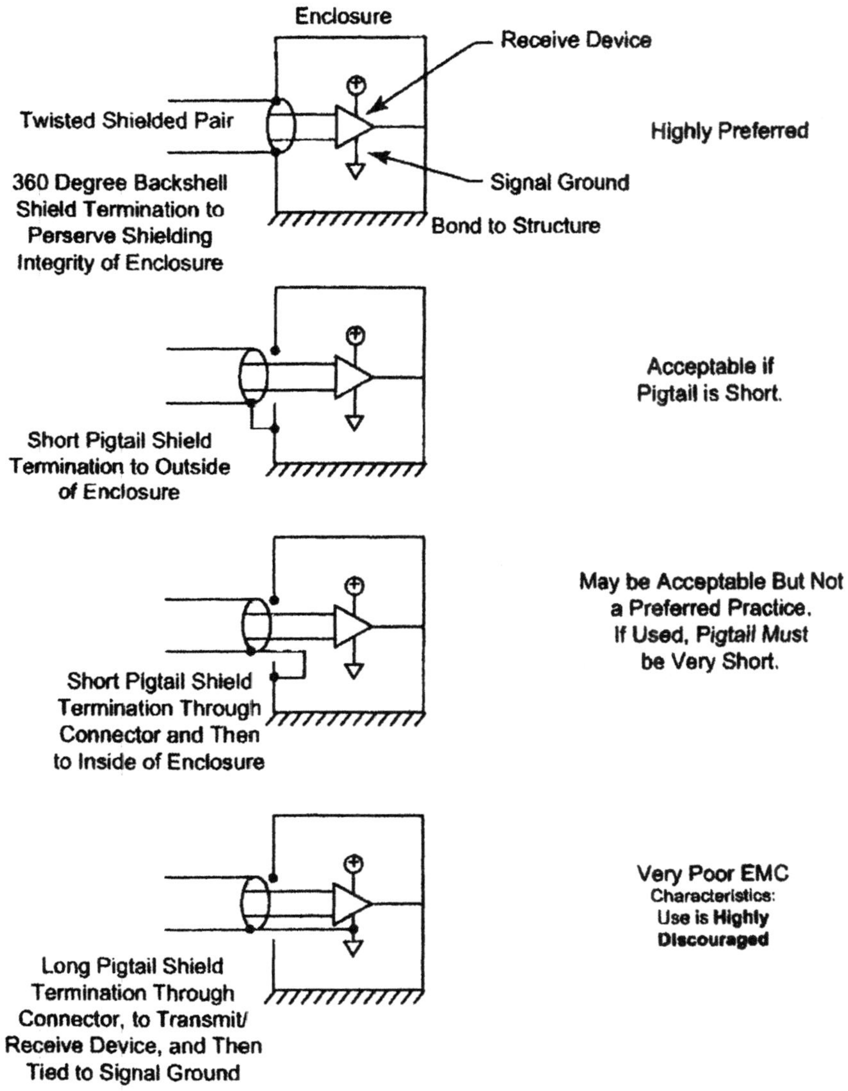

Figure 8.17. **Wire Terminations**

In some applications, double shielding of cables is required to prevent unwanted electromagnetic energy from entering the circuit. Figure 8.18 shows example schematics of how to ground double shielded cables. For some low-frequency, high-load-impedance circuits, grounding the shield at both ends causes low-frequency noise currents on the shield to couple into the circuit. Figure 8.19 is an example of a possible solution for this problem.

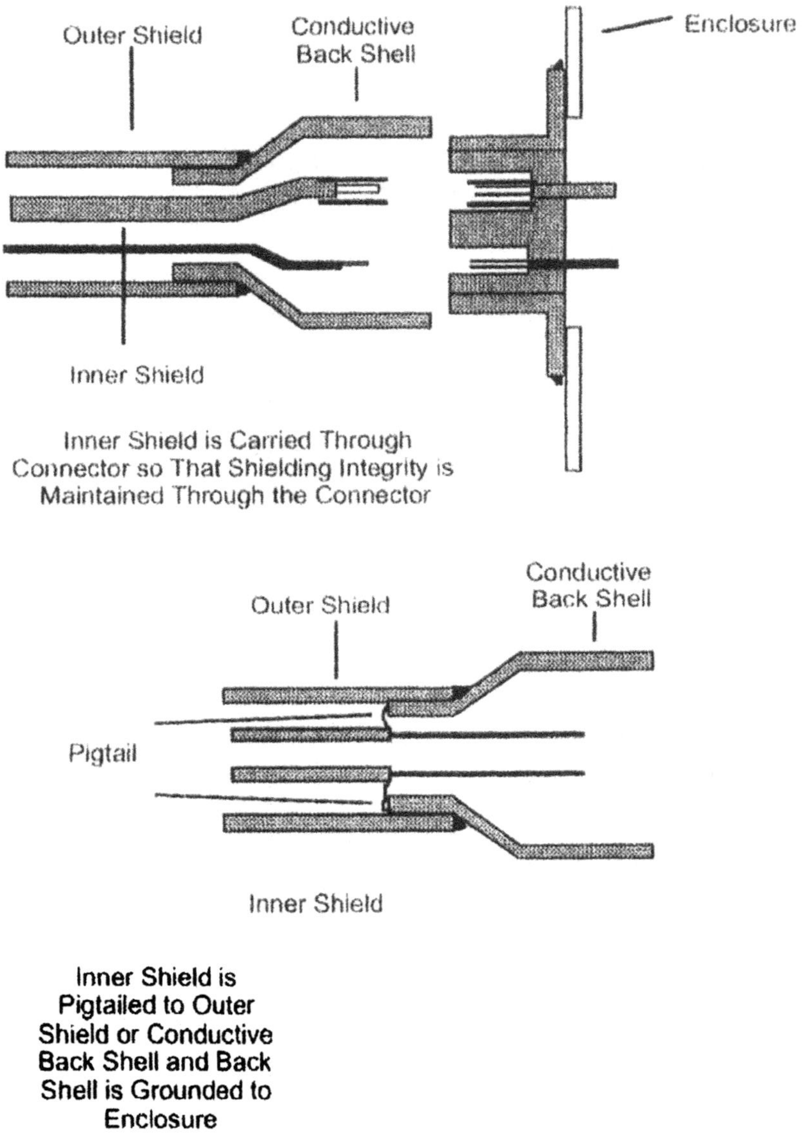

Figure 8.18. **Terminating Double Shielded Cables**

Outer and inner shields are grounded at one end and ungrounded at the other end. the two shields are isolated from each other at DC. At high frequencies, the capacitance between the inner and outer shield is such that the shield acts as if it were electrically grounded at both ends. This is effective against high frequency radiated fields.

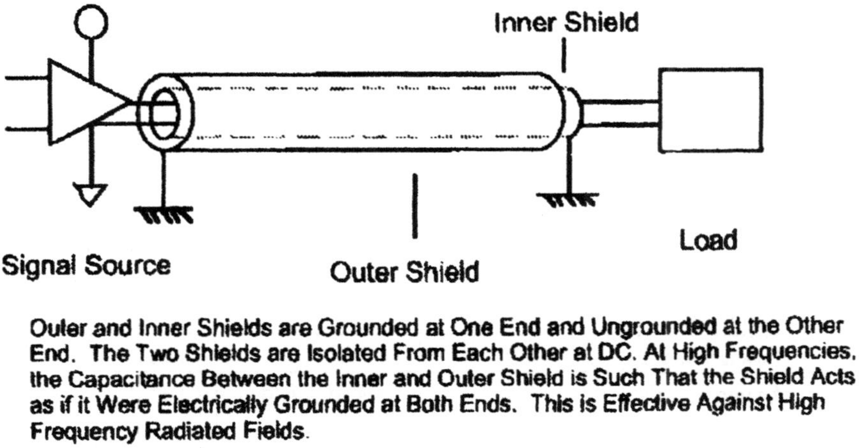

Figure 8.19. **Shielding Low Frequency, High Impedance Circuits**

8.2.11 Cable and Wiring Classes

Use of power and signal cables is prevalent on all spacecraft and payloads. These cables may act as both transmitting and receiving antennas for radiated EMI and conduits for conducted EMI. Because cables are usually routed to accommodate practical routing paths and equipment location, it is almost impossible to predict and quantify the EMI environment associated with these cables. One way of controlling EMI from cables and wiring is to separate cables and wiring into similar classes of voltage, frequency, and susceptibility levels.

Much guidance in this area can be obtained from practices developed for the NASA space shuttle, International Space Station *Alpha*, and U.S. military. These specifications have requirements or guidelines for wiring classification and separation. For example, the U.S. Air Force Systems Command Design Handbook 1-4 suggests the classification of wiring based on type of electrical power (ac or dc) and frequency susceptibility. Also, as a design goal, Design Handbook 1-4 suggests a minimum separation of 2 in (51 mm) between different wire classifications to prevent cable-to-cable coupling. NASA specifications for the Spacelab payloads and space station program (MSFC-SPEC-521B, Electromagnetic Compatibility Requirements on Payload Equipment and Subsystems) and (SSP 30242, Space Station Cable/Wire Design and Control Requirements for Electromagnetic Compatibility) contain requirements for cable classifications and separation. For programs in which such requirements are not supplied, Table 8.3 is a guide. The cables and cable bundles of different classifications should be separated by a minimum of 2 inches Figure 8.20 gives examples of the wire types called for in Table 8.3.

Table 8.3. **Suggested cable classifications.**

Signal Type; Rise, Fall Time *(tr, tf)*	**Voltage or Sensitivity Level**	**Wire Type**	**Circuit Class**
Power (ac, dc)	>6 V	Twisted	Class I
Analog Signals *tr, tf* > 10 *ms*	<6 V	Twisted Shielded	Class II
Analog Signals *tr, tf* > 10 *ms*	≤100 mV	Twisted Double Shielded	Class III
Analog Signals *tr, tf* < 10 *ms*	≤100 mV	Twisted Double Shielded	Class IV
Analog Signals *f* > 100 kHz, Digital Signals	A l l	Twisted Shielded, Coax	Class IV

Twisting the wire minimizes the loop area of the wire. This minimizes the amount of inductive noise coupling between the circuit and surrounding cabling. The number of twists per foot of cabling is limited by cable size; however, the greater the number of twists per foot, the smaller the loop area of the wire. Twisting will also increase the parasitic capacitance since the product of LC is a constant.

8.3 FILTER PLACEMENT

When all other attempts to correct EMC issues due to energy being conducted along a wiring harness have been considered and/or implemented, sometimes the only alternative is to use a filter network installed on the harness. Filters are not a preferred solution in the auto industry because of their additional weight, cost, and likelihood of not being re-installed after service procedures. Sometimes they may be the desired choice for perhaps very limited application or special vehicle packages.

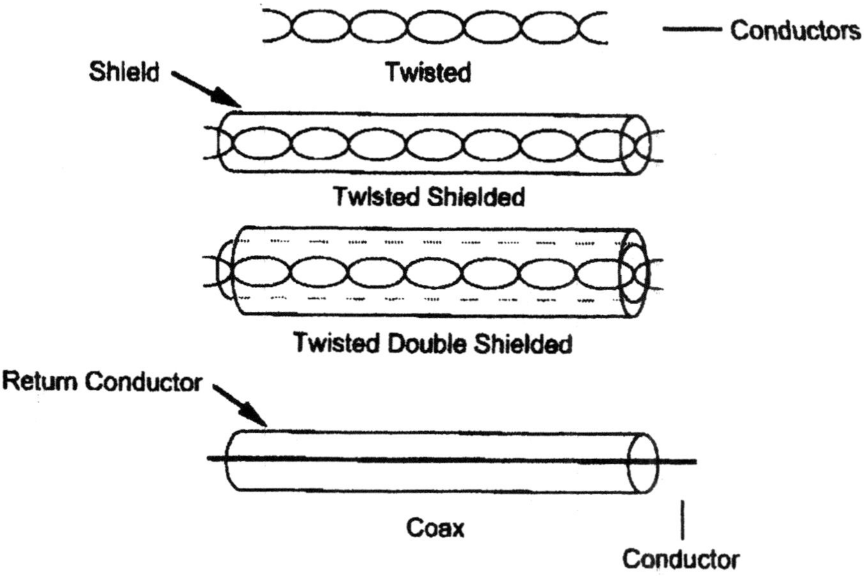

Figure 8.20. **Wire Types**

The important aspect to installing filters on a wiring harness is that the filter (or a ferrite-based suppression device, which is discussed elsewhere in this book) is to be located close to the source of the emission as shown in Figure 8.21. This will provide the maximum amount of suppression due to radiation from the cable (since the length will be minimized) as well as inserting a large impedance in the current flow. It is common to see filters being used on many other types of electronic devices, such as computer equipment and peripherals. In those applications, they may be a cost-effective solution. We are discussing their use here just to bring up the concept and point out that there may be some automotive applications where this is appropriate.

When considering the impact of and effect of wiring upon system level interaction, there is a "rule of thumb" that can be used to determine how to model the wiring. This is related to the physical attributes of the wiring, and is compared to the frequency of the signals that are of concern. These signals may also contain various amounts of noise.

The rule of thumb that can be used (based on empirical data,) suggests that as the physical length is much less than the wavelength (as it approaches 10 % of a wavelength and less), then the equivalent "lumped" elements can be substituted into the system model as parasitic components created by the wiring as shown in Figure 8.22.

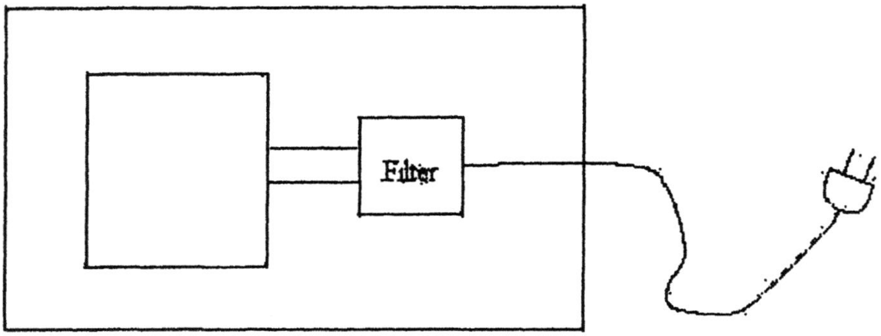

Figure 8.21. **Filter Placement**

This model can also be used to determine the parameters that are contributing to "cross talk" issues that may occurr.

We've also discussed the impact of conduction emissions upon system-level EMC issues. Conducted emissions can be difficult to work with in many problems, as sometimes the frequencies involved and the wiring that is associated with CE may change the problem into a RE issue. A good example of this is the CE that may be seen from an electronic device that is connected to the power grid via wiring. This condition is shown in Figure 8.23:

This is why many devices employ power line filters to minimize this coupling from the device to the power supply wiring. Otherwise, these power lines may essentially become large "antennas" that radiated energy from the CE within the device.

Sometimes a wiring harness has current flowing on it that is the noise current. By reducing this noise current, the amount of CE and possibly resulting RE can be reduced. A unique way to do this is by the incorporation of a high impedance path in the wiring harness. It is also desirable to leave the wiring undisturbed by not connecting additional components to the wires. One solution may be to use a device that increases the impedance of the circuit to the noise, yet does not significantly impact the desired signal.

These devices utilize ferrite materials and may be in a bead or a clamp configuration. The operation of these devices provides a high impedance path at the noise frequencies with little impact upon the intended signal.

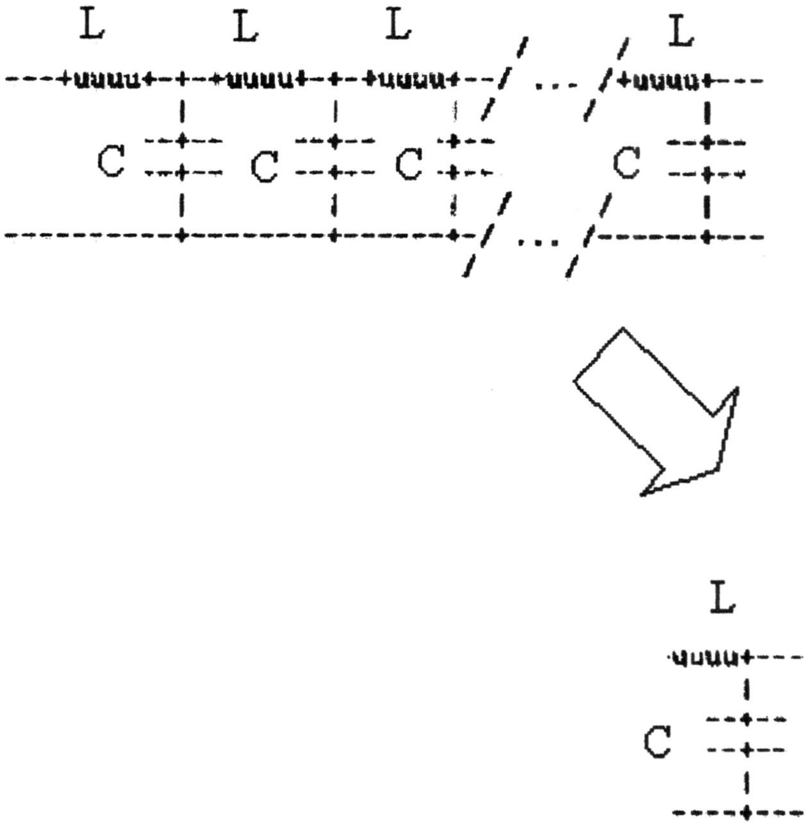

Figure 8.22. **Transmission Line Distributed and Lumped Elements**

Figure 8.24 is a photograph of several types of ferrite devices. These are available in many sizes and impedances.

It is important to place these devices near the sources of the emissions – to minimize the radiation due to CM current. These devices are less effective when further away from the source.

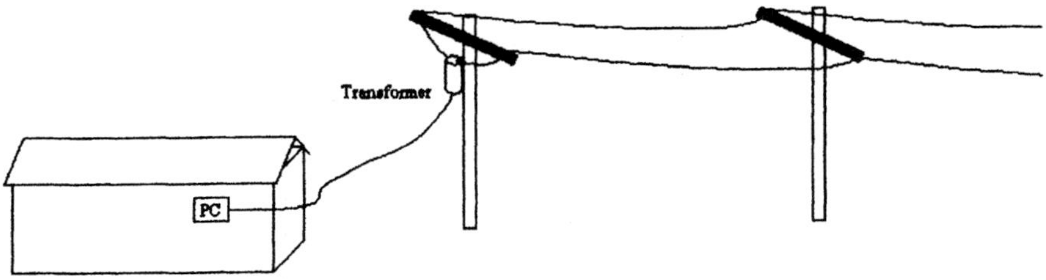

Figure 8.23. **Power Grid can Turn Conducted Emissions into Radiated Emissions**

Figure 8.24. **Ferrites**

The attenuation is due to the sum of the impedances. These impedances consist of the source impedance, the load impedance, and the additional impedance of the ferrite device. The attenuation is expressed in dB and is equal to the following:

Insertion loss or attenuation (dB) = 20 log (Zsource + Z load + Z ferrite / Zsource + Zload)

For example, if a circuit has the load and source Z equal to 50 ohms, and the impedance of the ferrite is 100 ohms, then the attenuation would be:

Attenuation (dB) = 20 log [(50 + 50 + 100)/(50 + 50)] = 6 dB

It can be seen that as the source and the load impedance increase, the amount of ferrite must also increase to provide the same amount of attenuation. Therefore, these techniques are primarily effective when the source and the load impedance are low.

Ferrite materials may be effective to reduce the magnitude of both DM and CM mode current. As was pointed out in other sections of the text, this is especially important for CM current, as a very small amount of CM current can have a major contribution to the RE of a system. Ferrites can also help reduce the current in a DM circuit, and this may be enough to allow the device to pass requirements or meet specifications. The impact of ferrites upon DM current contribution to RE can assist the other method of minimizing RE, reducing the effective loop area of the circuit. This method was discussed earlier.

The tradeoffs that must be considered are that the reduction of the current may result in functional issues. The effectiveness of ferrites would need to be evaluated in these applications. Another issue is that ferrites can be most effective in the early design stages, where it may be possible to configure the circuit to have low impedance to allow the use of ferrites to maximum advantage.

From a practical standpoint, as far as automotive system are concerned, it is not always practical to have ferrite devices installed on wiring harnesses to reduce the current levels. It is most applicable at the component level, within the component itself. The reason for this is that the ferrites are brittle devices and can be damaged by mechanical impact. Another aspect is that they may be difficult to reinstall after the vehicle is serviciced. Ferrite devices do play an important role in other types of electronics, specifically computers, where they can be made as part of an assembly, as shown in Figure 8.25.

In the automotive industry, the uses of ferrites do have their major role in aiding the diagnostic process.

Figure 8.25 **Ferrite on Computer Cable**

Let's look at the extreme example of conducted immunity or conducted transients. This extreme example would be a practical system that is set up with a very large potential between the earth or this system and the atmosphere. This may cause lightning to be attracted to the system. Lightning can be thought of as a severe case of conducted transients, because lightning has a significant amount of current flow associated with it and is very much like an arc discharge common in which some of the elements are destroyed in the process of the discharge! (One the authors has had personal experience with this phenomenon.)

8.4 COUPLING BETWEEN WIRES

This is a unique condition that is experienced in the auto industry that does not have many parallels in other industries – that of the extensive use of significant amounts of wiring to connect many components. This wiring must be as inexpensive as possible, yet be robust to extreme environmental stresses. Much of the time automotive EMC focuses on specific components, and neglects the impact of the wiring. One of the areas open for study right now is the amount of coupling that can be tolerated between systems. See Figure 8.26. Perhaps there can be some guidance obtained from the work that has been done in the NASA programs. NASA documents SSP30242E and SSP30243 discuss the metrics associated with crosstalk. What is interesting about the metrics is NASA states that the wires with potential crosstalk must have greater than 20 dB isolation (based on test results) and greater than 34 dB (based on analysis results). What does this means as far as actual values of voltage and/or current?

If we have two conductors parallel to each other, and conductor 1 carries one amp, the NASA requirements would require that the maximum current that could be induced in the other conductor would be the following:

$dB = 20 \log(10) (I_2/I_1)$ and $I_1 = 1$ amp

By solving for I_2, the maximum current that could be induced in the other conductor would be 20 mA, if this is verified by testing. If the amount were to be determined by analysis, the maximum value of I_2 would be 34 dB lower than the one amp of the current on the first conductor.

Traditional crosstalk work in EMC has been concerned with the relationship of three elements in the model. This model is similar to the basic EMC model, in that the crosstalk condition is due to the interaction of a source (sometimes called generator), a receptor (sometimes called a victim) and the path (the coupling that is taking place). (For a comprehensive study of this, see Dr. Clayton Paul's book *Introduction to EMC* – listed in references.) This model has additional complexity in the case of an automotive system because sometimes that length of the wiring harness is a significant portion of a wavelength of the noise that is being coupled. Under these conditions, it is necessary to use the distributed element model (or transmission line theory).

Understanding the inductance that is created by a loop of wire is important to know to determine the impact upon a system (which many people forget). Even a pulsed DC current can generate a few volts with only a few uH's of inductance – which is a typical amount in an automotive wiring harness.

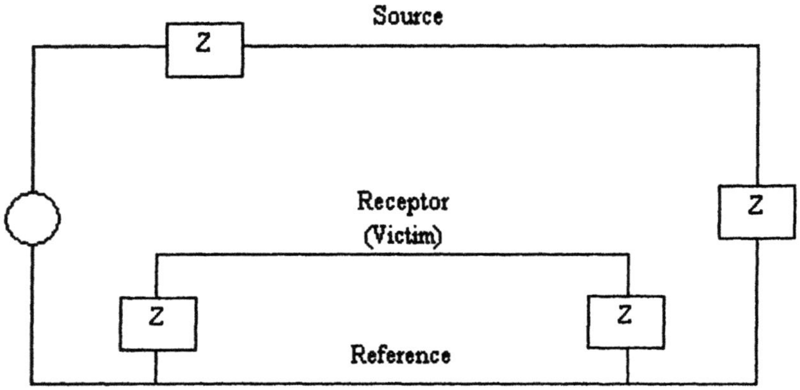

Figure 8.26. **Crosstalk**

8.5 GROUNDING AND PCB LAYOUT

The classical definition of a ground is "an equipotential point or plane which serves as a reference for a circuit or system". Unfortunately this definition is meaningless in the presence of ground current flow. Even where signal currents are negligible, induced ground currents due to environmental magnetic or electric fields will cause shifts in ground potential. This emphasizes current flow and the consequent need for low impedance, and is more appropriate when high frequencies are involved. It is important to remember that two physically separate "ground" points are not at the same potential unless no current is flowing between them.

Careful placement of ground connections goes a long way towards reducing the noise voltages that are developed across ground impedances. But in any non-trivial circuit, it is impractical to eliminate circulating ground currents entirely. The other aspect of ground design is to minimize the value of the ground impedance itself.

Ground impedance is dominated by inductance at frequencies higher than a few kHz, and the inductance of a pcb track or wire depends mainly on its length, given a "loop encloses" geometry. For example, a 10cm length of 0.5mm track appears as 60nH, while a 2cm length is 12nH. Paralleling tracks will reduce the inductance pro rata provided that they are separated enough to neutralize the effect of mutual inductance. The logical extension to paralleling ground tracks is to form the ground layout in a grid structure. This maximizes the number of different paths that ground return current can take and therefore minimizes the ground inductance for any given signal route. The limiting case of a gridded ground is when an infinite number of parallel paths are provided and the ground conductor is continuous, and it is then known as a ground plane. This is easy to realize at the PCB level on a multilayer board and offers the lowest possible ground path inductance, provided that it is not interrupted by other tracks. Note that the purpose of the ground plane is not to provide shielding but to give a low high-frequency ground impedance.

The important rule is that there should be no breaks in the plane layer underneath critical tracks (emissive or susceptible nodes) which would divert return current flow and hence increase the effective loop area (Figure 8.27); nor should these tracks be located near the edge of the plane. Thus it is necessary before starting the layout to identify which these critical tracks are. The criteria include ranking them in order of frequency and di/dt (for emissions) and in order of bandwidth, impedance and level (for analogue immunity) or whether or not they are latched (for digital immunity).

We've spent a great amount of space discussing the impact of wiring in an automotive system. Since we've done so, it should be clear why understanding the impact of wiring upon EMC issues in auto systems is important. In summary, the reasons are,

- Early systems and vehicles had few components to be connected
- Recent systems have increased wiring complexity
- Wiring will still be used for many years in the future
- Future electronic content of vehicles will increase greatly – with the advent of expanded uses of technology.

8.6 FERRITES

Ferrite is a common material for magnetic cores with relative permeabilities ranging from 40 to 10,000. Because of interwinding capacitance, multiple-turn cores have limited usefulness at higher frequencies. Hence, only ferrite beads with a single turn are considered here. Beads are very effective in limiting coupled energy to leads. A convenient property of ferrites is that the impedance is resistive above a given corner frequency. This implies that the coupled energy is dissipated and not reflected as heat. To obtain the required impedance, the general geometry of the bead should be longer. This is more effective than increasing the outer diameter.

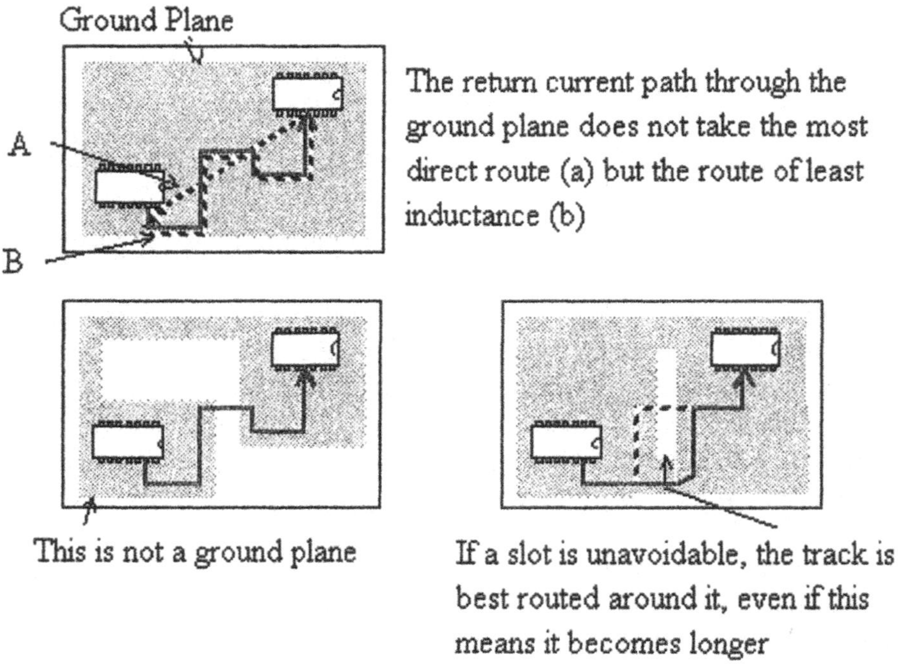

Figure 8.27. **The Ground Plane**

8.6.1 Ferrite Toroids

Large diameter toroids (1 to 2 inches) are used externally to the EUT to limit CM currents that cause radiation. These cores are typically used with 3 to 10 turns to increase the net impedance at lower frequencies. Again, the interwinding capacitance is in parallel with the inductor and, by shunting the inductor, limits the high-frequency performance.

8.6.2 Clamp-On Ferrites

At higher frequency ranges, acceptable results are often obtained with a single turn. Many manufacturers have developed split cores with plastic retaining housings to clamp the ferrite on a cable. Geometries are available for coaxial or round cables as well as ribbon cables.

Theory of cable shielding at medium frequency and beyond (300 kHz +) is quite different than for instrumentation-type shields. The shield must not only have a low impedance to ground (at both ends) but also provide for physical separation of currents interior to the shield versus exterior ones. The shield acts as a "container" of EMI; in fact, it is an extension of the equipment enclosure. The shield is analogous to a tunnel or enclosed walkway between two buildings; its purpose is to preserve and extend the environment of the two buildings in the space between them. The cable must not leak and allow mixing of two separate environments. Since currents flow on the inside of the cable shield is due to EUT generated noise, the shield termination must prevent those currents from finding a way to the outside. A 360° peripheral termination accomplishes this separation. The skin depth of the shield material (presumably resulting in low shield transfer impedance) provides the separation function on the cable shield.

8.7 ATTENUATING COMMON MODE CURRENTS ON UNSHIELDED CABLES

If an unshielded cable is determined to be the source of RE, via use of a current probe, the parasitic coupling of EMI to the cable must be attenuated. A current probe measures only the net current flowing through its window. If the cable is designed correctly, the intentional signal is not measured by the probe (both signal and return lines pass through the probe window). Any CE measured in-band to the RE failure are unintentionally present on the cable and, therefore, are filtered or otherwise removed from the cable without affecting the intentional current. Techniques for achieving this are CM chokes implemented as RF beads and line-to-ground capacitance providing the capacitors do not load the intended signal.

If these after-the-fact bandaid approaches do not work, it may be necessary to redesign the cable interface at the equipment enclosure. A bandaid approach is to shield the cable. Another approach is to determine how the EMI is coupled to the cable and to perform an isolation at the circuit level (source suppression). This might involve redesigning the PCB layout, or choosing interface circuitry which inherently provides isolation (opto-isolators, transformer coupling, etc.).

8.8 HIGHER-FREQUENCY EMISSIONS

When the frequency of the emission is such that the wavelength approximates the EUT enclosure dimensions, seams and apertures in the enclosure are suspected sources of emissions. A general rule is seams and apertures are shorter than one-tenth wavelength in order to avoid leakage.

Chapter 9

Automobile Electrical/Electronics Systems

9.1 VEHICLE GENERATED RADIATED EMISSIONS

Automotive systems can be sources of radiated emissions that may affect on-board electronics, or interfere with electronic systems on adjacent vehicles or devices along the roadside, such as televisions, radios, etc. Active electronic devices generate radiated emissions during their normal operation because of switching operations, logic gates, PWM (pulse-width-modulated) control signals, etc. Even the non-solid-state components, such as mechanical switches, horns, relays, other inductive devices, and spark plugs can also radiate. Many of these devices have been used since the early days of the automotive industry, and still require attention to ensure that measures are taken to prevent problems.

The automotive industry has classified emissions from electrical and electronic systems in two ways. They are classified BB (broadband) or NB (narrowband). Several methods can be used to discriminate between these types of emissions. The reader is encouraged to review the material that is referenced at the end of this chapter. These classifications may vary as used in different industries.

The auto industry has a straightforward approach to classification of the emissions (also called noise), which as follows:

If the emission is created by an "arc" or "spark", it is classified as broadband (BB) noise.

This means that all other devices are classified as "narrowband" (NB) emissions.

While this may not be the most rigorous method of classification, it is practical and has worked reasonably well for many years.

BB noise is emissions that occupies a wider frequency spectrum than the bandwidth of the receiver in use. This means that it is not possible to tune out of the emissions and tune to a frequency without emissions. However, with NB emissions, there may be frequencies that are occupied by the emissions sources and other frequencies that are not occupied by the noise. By looking at Figure 9.1, you can understand how these terms relate to the actual physical process.

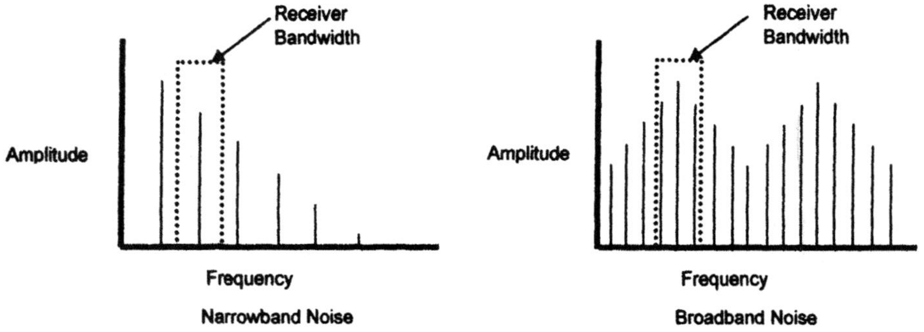

Figure 9.1 **Narrowband Noise and Broadband Noise**

Imagine that a sliding receiver "window" defines what is heard in the receiver. By moving this window across the BB noise in Figure 9.1, you can see that since the emissions change as a function of time, there are always a significant amount of noise emissions in the receiver's bandwidth. Now, if the receiver's window is moved across the NB noise spectrum, there may be a setting where no noise falls within the receiver's pass band.

9.2 BANDWIDTH RELATES TO "SELECTIVITY"

Bandwidth relates to how selective the receiver is. Consider the following examples of the various audio bandwidths that are in use today.

- The BW of a telephone audio circuit is between 3 and 6 kHz.
- The bandwidth of the typical AM broadcast receiver is 7 to 8 kHz

- FM broadcast stations, used for high-fidelity transmissions of sound, have the greatest bandwidth at 100 to 150 kHz.

Figure 9.2 shows the BW of an AM broadcast receiver receiving one of several AM radio signals. It can be seen that there will be only one signal that is received at a time for each of the AM frequencies.

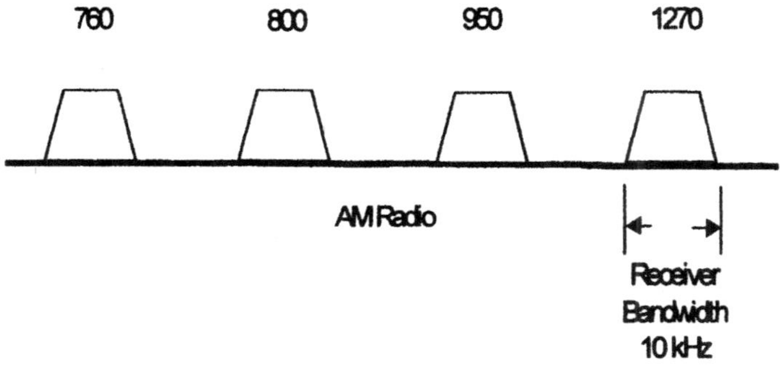

Figure 9.2 **An AM Radio Receiver can Select the Desired Station**

Figure 9.3 shows that if the BW of the receivers is wide (not selective enough), then for signals that are close to the desired frequency, these adjacent signals will also enter the receiver.

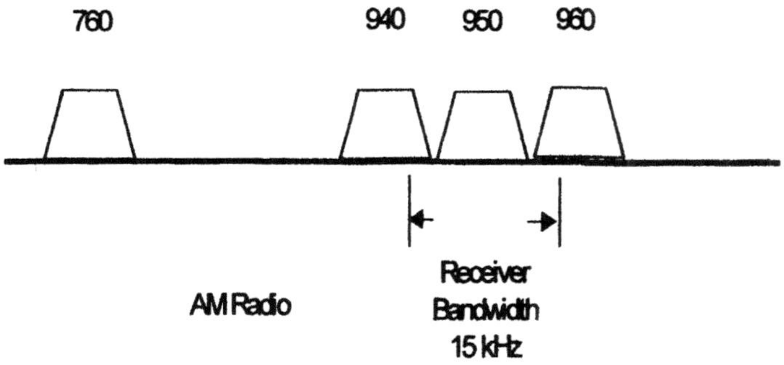

Figure 9.3 **Inadequate Selectivity Permits Adjacent Signals to Interfere with the Desired Signal**

9.3 BROADBAND NOISE

9.3.1 Motor Noise

It is easy to observe the effects of BB noise from an electric motor. There are examples in everyday applications, such as vacuum cleaners or mixers being operated in a home. It is not even necessary to have high power for the noise generation. Many "cordless" battery operated devices can emit enough energy to generate BB noise. There are many types/applications of motors in many environments, and of course, automotive systems also use many motors. These are used to adjust seat positions, open and close windows, adjust mirrors, in HVAC operation, and in powertrain control functions.

Another issue with motor noise that is that in addition to the energy being radiated from the motor itself, the conducted noise along the power feed from the battery to the motor can contribute to radiated emissions. This is an issue in automotive systems because of the wiring that connects all loads to a common battery power supply. The harness can look like a low impedance path for the noise energy. As discussed in other sections of this book, current takes the path of least impedance (the phrase "least resistance" is only a special case for DC). Knowing that noise current takes the path of least impedance this will improve our ability to analyze EMC problems. When a motor is connected to its power supply wiring, in addition to being a source and return for the power, the wires also can act as a path for noise current or, depending on the frequencies and harness length involved, as an efficient antenna.

9.3.2 Ignition Noise

Figure 9.4 shows a representative internal combustion spark ignited (SI) engine ignition system. Similar systems have been used since the beginning of the auto industry, and though some aspects of the technology have changed, the basic elements remain unchanged.

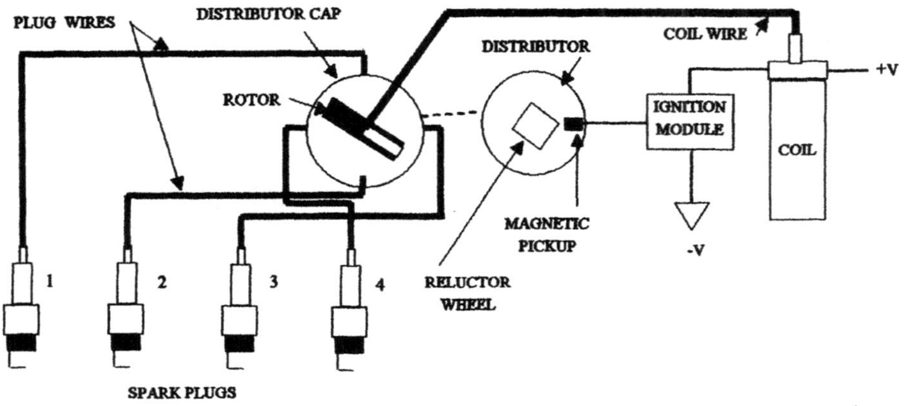

Figure 9.4 **Spark Ignition System**

There must be some method to store the energy used to generate the spark. In this system, an inductive element (ignition coil) stores energy and then releases it at the correct time during engine operation. This timing information comes from some type of engine camshaft or crankshaft position sensor. In early auto systems, this position information was directly linked to the opening and closing of the points, which acted like a switch. The next requirement for an SI engine is to have a spark generated by this energy. Reviewing the information regarding inductors (from Chapter 4) – there is a voltage that is generated by the rate of change of the (primary in this case) current. This equation is:

$$V = - L \times dI/dt$$

Equation **9.1**

L = the inductance (in Henry's)

dI or "delta I" = difference in the current from its maximum to its minimum values

dt or "delta t" = the amount of time it takes to go from the maximum to the minimum (or vice versa) of the current

For the case of an ignition coil, typical values of L are in the millihenrys, and the total current change is 5 to 10 amps, accomplished in a few microseconds. Calculating the potential voltage at this discharge then yields tens of kV that are available to start the combustion process. As this high voltage is generated, it is transferred by the distributor (a high voltage switch) via the secondary wires to the spark plug's gap, where the discharge takes place. This is the source of the noise. It is what many ignition systems have in common, no matter how they are constructed; they all utilize a coil, spark plugs, and a method to control the primary current. Other types of ignition systems, such as capacitive discharge though those are not as common, perform essentially the same function.

In earlier model vehicles, there were actually two spark discharges that could occur, one in the distributor, which was a mechanical switch used to route the HV to the correct plug, and another at the plug itself. Recall that the amount of noise generation is based upon the dI/dt of the coil, and this maximum value occurs during the spark initiation region (discussed in Chapter 4).

If we look at the spectrum analyzer plot of noise from an ignition system in Figure 9.5, we will see the familiar profile of BB noise discussed earlier. Figure 9.5.a shows the emissions as a function of time. Figure 9.5.b shows the resulting envelope over time. Placing the spectrum analyzer in "peak hold" mode, as time continues, and amplitudes and specific frequency content vary, the display will be filled in with the envelope of the noise at all frequencies on the display. Given enough time, the display will show a solid band of emissions from start frequency to stop frequency.

9.3.3 SCR Noise

Another example of BB noise generation is from silicon controlled rectifier (SCR) controlled devices. Using a SCR to control the voltage to an AC operated device means that the sine wave is truncated at some point in its waveform. By doing this, the waveform starts to look like a portion of a square wave, with (as discussed earlier) harmonic content. This can be heard by listening to an AM radio while an incandescent lamp dimmer is used to control a lighting circuit in the same room. Some of the newer light dimmers incorporate coils to "choke off" noise current from the house wiring. See Figure 9.6.

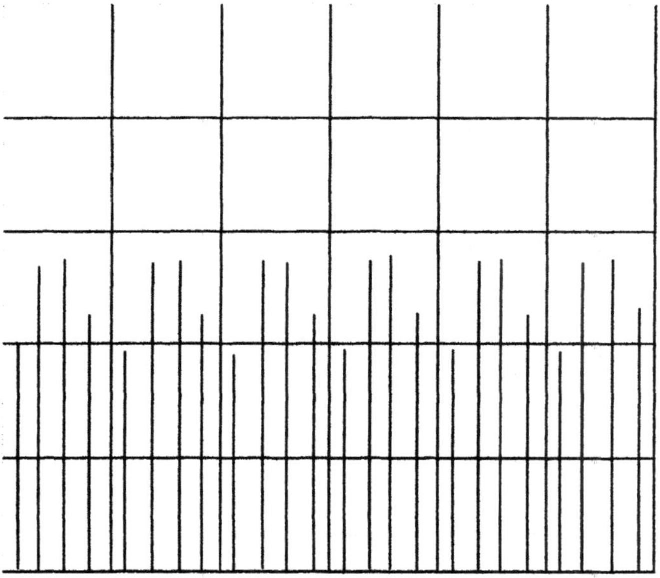

Figure 9.5.a. . **Ignition Noise as a Function of Time**

Figure 9.5.b **Ignition Noise Envelope**

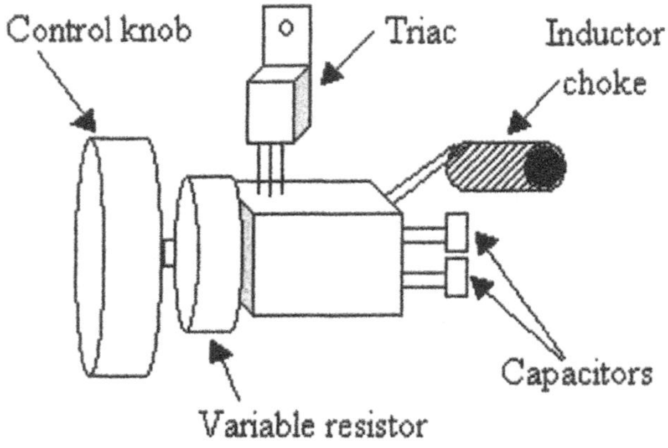

Figure 9.6. **Light Dimmer Inductor Noise Suppression Method**

The important point from an automotive aspect is that we can have a number of devices that either emit noise themselves and/or emit noise current on the wiring, and this noise can be radiated from a vehicle and picked up some distance away.

To review – the noise that is emitted from motors is BB. It appears on a spectrum display which will show many noise pulses within the instrument resolution bandwidth. These noise pulses will extend over a wide range of frequencies. Although most of the noise is generated by the motor brushes, even the "brush-less" motors can emit noise because of the switching of the current. The impact of electric motor noise on auto entertainment systems depends on the type of radio signal modulation. On AM radios, the noise will be heard as a "pop" or "click". In FM radios, the noise may not sound like noise, but may reduce the radio's sensitivity. The noise reduces the input SNR (signal-to-noise ratio). The reduced SNR will reduce the AM radio's output SNR. The FM radio may simply "quiet" at a higher input signal level, reducing the range over which a particular station may be received.

One of the characteristics of BB noise is that it is not possible to "tune out of" the noise. This can be seen with a simple demonstration using an AM radio and a source of BB emissions. These emissions can be from any type of high voltage discharge device (with a continuous discharge) such as a neon sign, a fluorescent light, or a motor. Tune into a station on the AM band, turn on the device, and note the interference. Tune the radio across the band and note how the noise still occurs at any frequency. This demonstration shows that there is a continuous band of energy that is being emitted across all frequencies.

9.3.4 Overview of BB Noise Sources

In summary, let's look at the characteristics and sources of BB noise. There are many systems and components in the automobile that generate BB noise. Some of these components and technology date from the early days of the industry, and others are recent technology. The early components include spark plugs, ignition coils, switches, and motors. These can have EMC (mainly EMI) issues associated with them.

In order to address emissions from SI ignition systems, it may be easier to work at the component level, rather than at the vehicle level. For example, the radiated emissions level from spark plug RFI was reduced in the 1950s to achieve compatibility with radio receivers, and these techniques are still relevant today (see references).

9.4 NARROWBAND NOISE

Let's review the characteristics of NB noise at this point. Recall from earlier discussions that we classify noise as NB if its bandwidth is less than the affected receiver's RBW. This means that there may be only one or a few frequencies emitted from the device that will interfere with a radio receiver. The nature of NB emissions is that the frequencies at which the noise is observed and the amplitude of the noise tend to be stable over time.

9.4.1 Microprocessors and narrowband noise

The following is an example of how the emissions from a typical microprocessor source could occur, as result of harmonics of the fundamental frequency that are emitted from the microprocessor's operation. If we look closely at this, we will see that there is a very repeatable pattern that occurs with the emissions. This very repeatable pattern occurs at multiples of the clock frequency. What is significant about this type of emissions is that the emissions will stay constant at the frequency and relatively constant with respect to amplitude overtime, unlike broadband emissions where both the amplitude and the frequency will vary. This is referred to as a "comb like" appearance, because the emission pattern looks like an inverted comb.

One of the major sources of NB emissions in today's automobiles is the microprocessor. In order to operate a digital microprocessor; we require clock and logic signals of sufficient amplitude. Secondly, microprocessors are being added in greater numbers to more vehicle modules. Lastly, designers of digital circuits prefer short pulse rise and fall times to minimize timing uncertainty and to reduce device heat dissipation. If we look at the spectral content emitted from microprocessor operation, it may resemble Figure 9.7, with the emissions at the harmonics of the clock fundamental frequency. Many of you may recall you learned that only "odd harmonics" would exist in a square wave. This is true ONLY for EXACTLY 50 % duty cycle signals, which seldom occur in the real world!

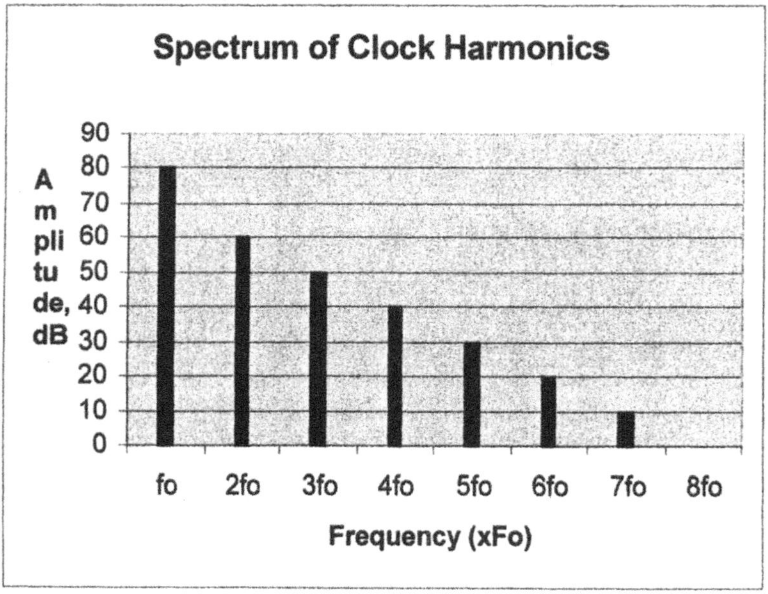

Figure 9.7 **Frequency Spectrum of Microprocessor Emissions**

Microprocessors are not the only source of the NB emission. Many other devices, such as power transistors, PWM (pulse width modulated) speed controls and switching transistors, can generate NB emissions.

We have discussed SI ignition systems and motors as an example of BB noise sources. We can also give a demonstration of NB noise from a microprocessor system. If we place a portable AM radio near a source of switched logic, we should hear the effect of the radiated emissions on radio reception. In this case, we will use the emissions from a novelty device (key chain) to demonstrate the process. There are many items with digital logic being switched at high rates. By activating the device and placing it near the radio, we will hear the effect of the logic switching. Another potential source would be the "musical greeting cards', as they employ similar electronics. If one of these cards is placed near an AM radio, the emissions from the card are picked up by the radio and can be heard.

9.4.2 Generation of narrowband interference

NB interference can be demonstrated very easily. Using a portable FM broadcast receiver, tune to a station in the FM band. Create an emission at that same frequency using an NB source (with a stable oscillator, or a RF signal generator). If the signal strength from this test emission is greater than the signal strength from the received station, then the received station will be blocked out by the emission from the test source. Note that, while this particular station would be blocked out, the radio would appear to function normally on all the other stations. This can be very confusing from a diagnostic process standpoint as it will look like the receiver and the system are operating normally except at that one particular frequency. This may also be accomplished by using a second FM radio tuned 10.7 MHz below the receive frequency of the first radio.

9.4.3 Narrowband radiate emissions case study

Just how much concern is radiated emissions from microprocessor based devices today?

Let's look at an actual case of NB radiated emissions on a vehicle. This shows what can happen when unanticipated radiated emissions fall on a frequency that is being utilized for vehicle operation.

The problem occurred on a transport truck. The two-way radio used to communicate with the dispatcher used frequencies in the VHF band. The radios could not be utilized because radiated emissions from the vehicle were on the communication channel frequency. To analyze the path that the emissions coupled to the radio, a series of tests was performed. By disconnecting the vehicle power from the radio and powering the radio from a battery, it was determined that the noise was not being conducted through the power leads or other wiring. By disconnecting the antenna cable from the radio, it was determined that the noise was being radiated from the engine control system to the antenna for the radio. (This process can be very helpful in diagnosing vehicle radio problems).

Several solutions were tried and the one that was finally selected was changing the frequency of the microprocessor clock reference oscillator crystal. Since the crystal fundamental frequency is multiplied many times to generate the VHF band harmonic, a very small change in the fundamental frequency can yield a significant change at the frequency of concern. This solution is shown in the block diagram in Figure 9.8. The diagram shows the generation of the emission at the communication frequency and how it created the problem. The block diagram in Figure 9.9 shows the effect of moving the fundamental frequency by a very small amount, and how because of the frequency multiplication, the emission moved from the frequency of concern to another frequency.

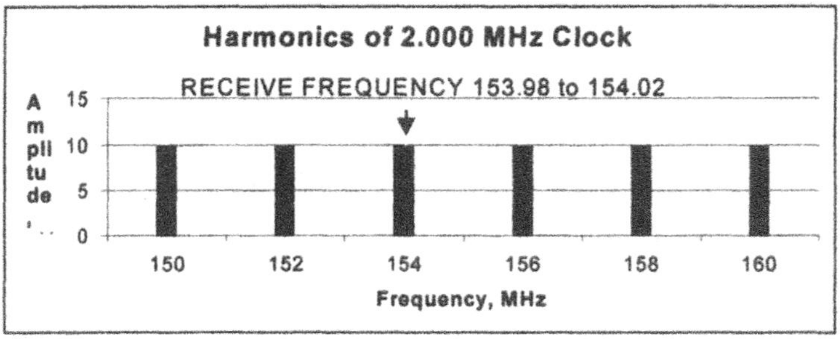

Figure 9.8 **Original Crystal Causes Interference**

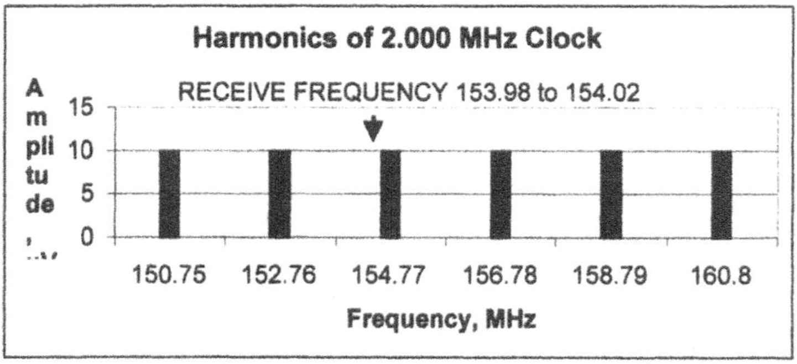

Figure 9.9 **Crystal With Slightly Different Frequency Eliminates Interference**

9.4.4 Impact of narrowband noise

In summary, NB noise sources may affect only specific frequencies, which can result in receivers appearing to function normally. However NB noise may often be addressed in the component design process by obtaining data early in the design or development process. By selecting a "quiet" microprocessor, and designing the circuitry to minimize NB emissions by increasing the impedance or providing "shunt path" for the emissions, radiated emissions problems may be minimized

Again, there are three characteristics of narrowband noise sources:

- NB noise may only affect specific frequencies.
- Receivers can appear function almost normally in the presence of NB noise.
- NB emissions can be addressed in component design process.

9.5 SIGNAL CHARACTERISTICS

We may see two different types of signals in digital circuits and systems.

The first one is called "deterministic", a signal whose behavior is precisely known. Examples would be sine-wave and digital clock signals, where the signal state is a predetermined relationship with time. For example, at time one the logic level may be at zero, and at time two it may be at logic level high. And at time three it would be again at logic level zero. This is shown in Figure 9.10. This is a predictable signal as function of time, for we know that at any point in time it is possible to predict the signal amplitude.

Figure 9.10 **A digital Clock Signal as an Example of a Deterministic Signal**

The second signal type is non-deterministic or random, a signal whose behavior is not known, and can only be described statistically. It is not possible to predict what the value of the signal is as a function of time. Digital data is an example. This is shown in Figure 9.11.

Figure 9.11 **Digital Data as an Example of a Non-Deterministic or Random Signal**

Another way to classify signals is periodic or non-periodic. A periodic signal is repetitive in the time domain such as a clock signal, square wave, or sine wave. Non- periodic signals are not repetitive in the time domain. They are called energy signals because their total energy is finite.

Let's review the amplitude vs. time for a square wave signal. If we look at Figure 9.12, we note that the amplitude of a square wave signal is constant and varies from zero to its maximum value, as a function of time. The frequency of the fundamental signal can be expressed as the following, where the frequency is equivalent to the reciprocal of 1/T.

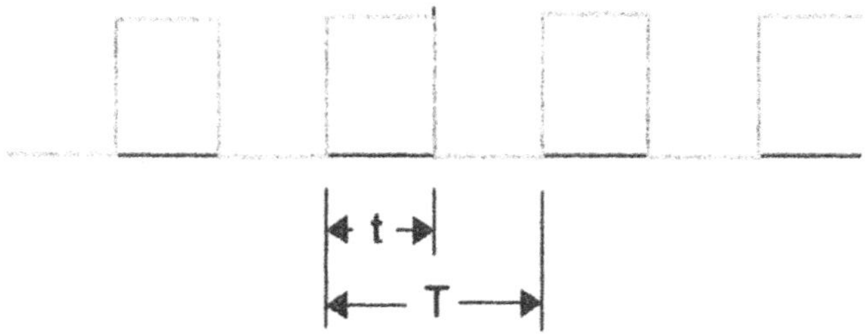

Figure 9.12 **Square Wave**

For a "50% duty cycle" square wave, the time, T, which corresponds to the period, is equal to two times t, the pulse "on" time. If we now express the average amplitude as being equal to the maximum amplitude times the amount of time that the signal is at the maximum amplitude and divided by the the total amount of time, that is equal to the maximum amplitude divided by two. The average amplitude for a 50 percent duty cycle square wave is half the maximum amplitude.

Let's look more closely at the square wave. A perfect square wave would have zero rise and fall time. In actuality a finite amount time is needed for the signal to rise and fall because of physical limitations. It takes a finite amount of time for the voltage level 1 to increase to its maximum from its minimum level. This referred to as the rise time. Similarly there is a characteristic called fall time. It is industry practice to express the rise time as tr., and the fall time as tf. It should be obvious at this point that both the rise time and the fall time must be greater than zero, as logic states cannot change instantaneously when dealing with real world components.

An additional characteristic that we referred to with regard to clock signals is the duty cycle, which means the proportion of time that the signal is at its maximum level as a function of the overall period T. For example a 50 percent duty cycle signal means that if the total period is T, then the "on" time T1 plus the "off" time T2. Must be equal, and their sum must equal T.

A simple demonstration of the concept of the harmonic content of a square wave can be performed with a function generator. Set the generator to a square wave output at a frequency of approximately 200 kHz (lower end of the long wave broadcast band) and prepare a communication receiver to receive signals at multiples of the 200 kHz signal. First set the receiver to 200 kHz, with the function generator output set to a square wave. Then switch the generator output to a sine wave. Note any difference in received signal. Next, set the receiver to 400 kHz, and switch the function generator to a square wave; note the received signal, then change the function generator output to a sine wave. With the square wave, the communication to receiver should receive a signal. With the sine wave, the harmonic at 400 kHz will not be present. Continue to set the receiver to successive harmonics (600 kHz, 800 kHz, and so forth). Each time change the output from the generator from a square wave to a sine wave and observe the received signal. If the generator accurately generates a 50 percent duty cycle signal each time the receiver is tuned to odd harmonics of the fundamental frequency, there will be a received signal which is a component of the square wave. Also note that each time the generator set to a 200 kHz sine wave that no signal will be heard on any multiple of the fundamental frequency. This demonstration of the relationship between a square wave and its harmonics can be useful to the understanding of harmonic emissions from logic devices.

9.6 RE DIFFERENCES BETWEEN "IDENTICAL" COMPONENTS

An interesting experiment has been performed to document emissions from a microprocessor assembly. The test used a microprocessor mounted on a circuit board which was termed the device under test. The emissions from the DUT were measured using a small loop antenna, shown Figure

from a microprocessor assembly. The test used a microprocessor mounted on a circuit board which was termed the device under test. The emissions from the DUT were measured using a small loop antenna, shown Figure 9.13.

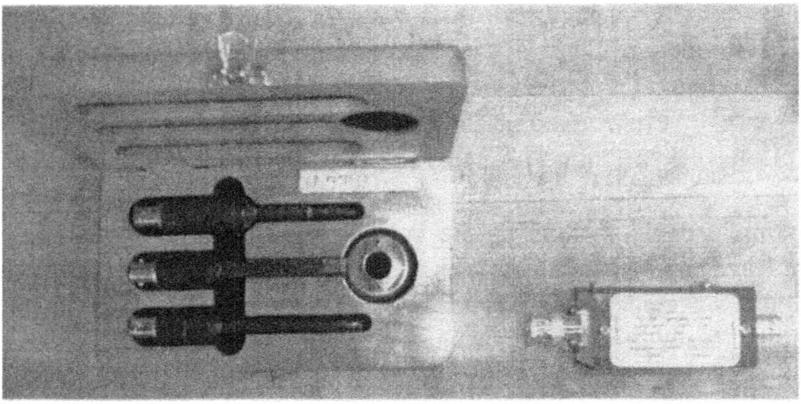

Figure 9.13 **Near Field Probes Used to Measure Radiated Emissions From A Microprocessor**

The data for the first microprocessor is shown in Figure 9.14. The horizontal axis spanned the frequency of the emissions and the vertical axis indicates the level of the emissions at each data point.

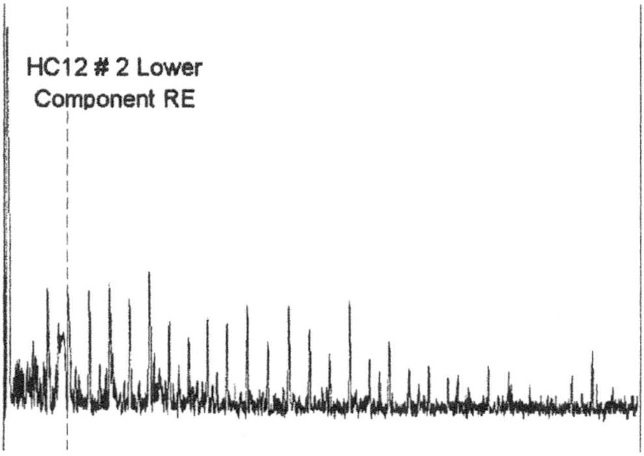

Figure 9.14 **Component Radiated Emissions From First Microprocessor**

The test was then performed with a microprocessor on a second board. Data is shown in Figure 9.15. The horizontal axis spanned the frequency of the emissions. Again, the vertical axis indicates the levels of the emissions at each data point. Note that at each frequency the emissions were different This is interesting because both these devices are similar and interchangeable from a functional standpoint. Yet we see that there are different radiated emissions levels from the devices themselves.

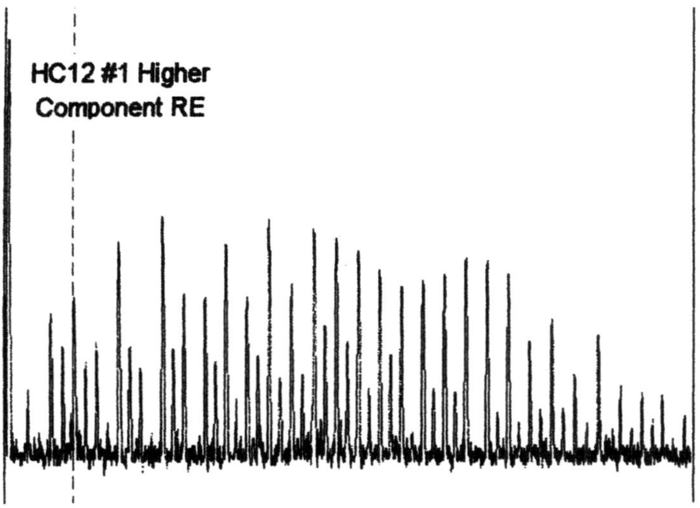

Figure 9.15 **Component Radiated Emissions From Second Microprocessor**

9.7 VEHICLE RADIATED EMISSIONS TEST

How would the results of this component test translate to vehicle radiated emission levels? Figure 9.16 shows the test setup for the vehicle in which each of the two different microprocessors could be evaluated. This test will measure the radiated emissions from the vehicle on a spectrum analyzer. There will be a high degree of correlation between radiated emissions seen at the vehicle level and those observed at the component level. Figures 9.17 and 9.18show that component level EMC measurements can be used to flag potential vehicle problems early in a vehicle program.

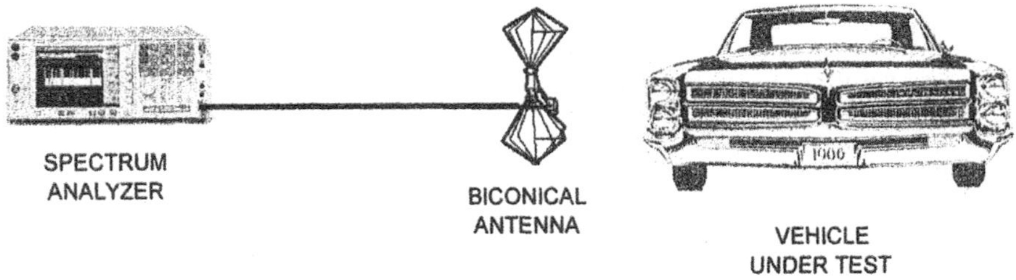

Figure 9.16 **Vehicle Level Evaluation of Microprocessor RE**

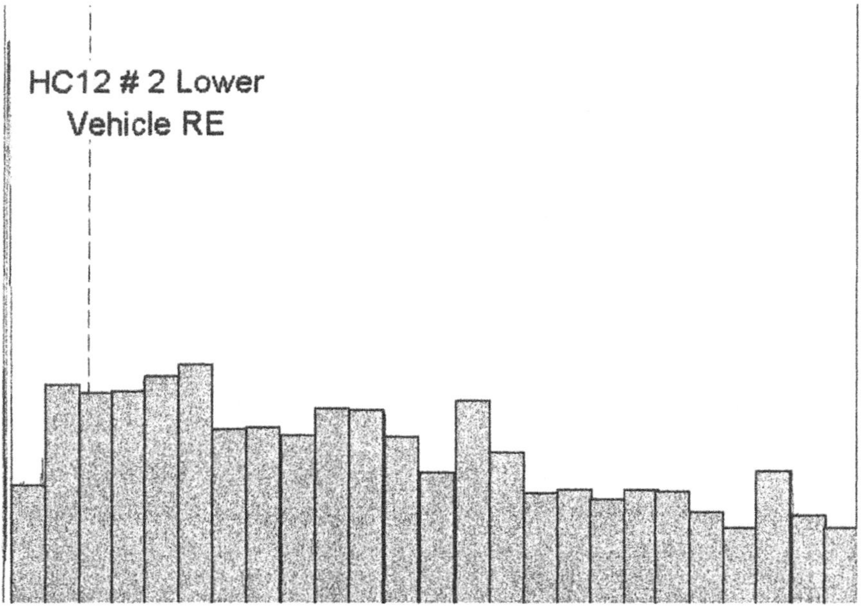

Figure 9.17 **Vehicle Radiated Emissions From "First" Microprocessor**

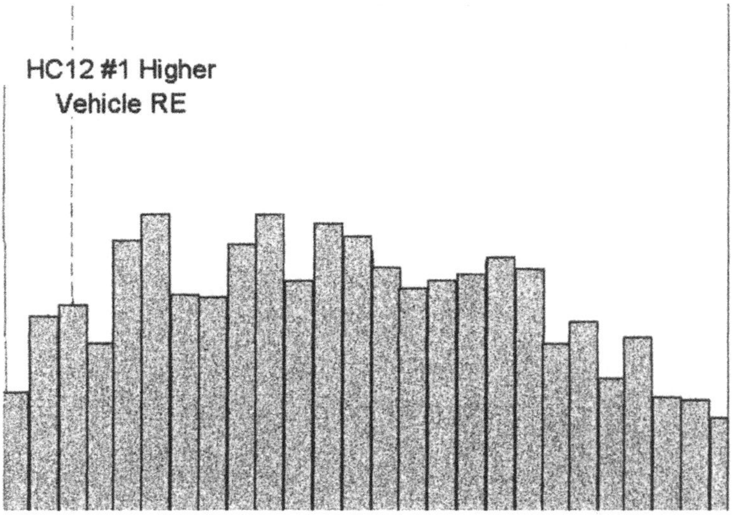

Figure 9.18 **Vehicle Radiated Emissions From "Second" Microprocessor**

9.8 SUMMARY

In summary, the data appear to show that a measurement can be performed to determine the relative EMC performance of components and then the data utilized to predict vehicle level performance. This tool would be utilized in the design and development process for new components to predict their expected contribution to overall system level performance.

9.9 DIGITAL SYSTEM DESIGN

Another consideration in designing real systems is that the implementation of those systems must include the limiting characteristics of actual electronic components like digital electronic circuit logic. Trade-offs exist using real systems. A refresher of digital system design and signal content will help.

Recall the Fourier series expansion:

$$f(x) = \frac{a_0}{2} + \sum_{k=1}^{\infty} a_k \cos kx + b_k \sin kx.$$

This equation indicates that, in theory, the harmonic content of the series goes to infinity. Clearly from a real system standpoint it is impossible for the signal to extend to infinity, because there has to be some limit to the practical amount of energy we are concerned with.

Essentially there are two points that the spectrum of the energy content of the harmonics is of interest. The first one is the fundamental frequency (which we will call fo.) The second is the "bandwidth" of the system. The energy content decreases at approximately 20 dB per frequency decade (a frequency ratio of ten to one.) For example if we have a fundamental at 2 MHz, the frequency one decade higher would be 20 MHz.

We need to define the term “bandwidth” with regard digital systems. The bandwidth is determined by the rise time (tr). The equation for the bandwidth is as follows:

BW = 1/(pi*tr)

As can seen from this equation the bandwidth is inversely related to the length of the rise time. Thus, as the rise time decreases the bandwidth increases. This is a fundamental concept important to EMC issues. The bandwidth point defines the frequency at which the energy content changes from a rolloff of 20 dB per decade to 40 dB per decade. This is shown in Figure 9.19, which shows data obtained with digital devices by adjusting the rise time of the signal. As a rise time decreases, the harmonic content increases for a given portion of the spectrum. The Figure shows the relationship between the fundamental frequency, the bandwidth and the rise time of the signal, and the roll-off of the energy.

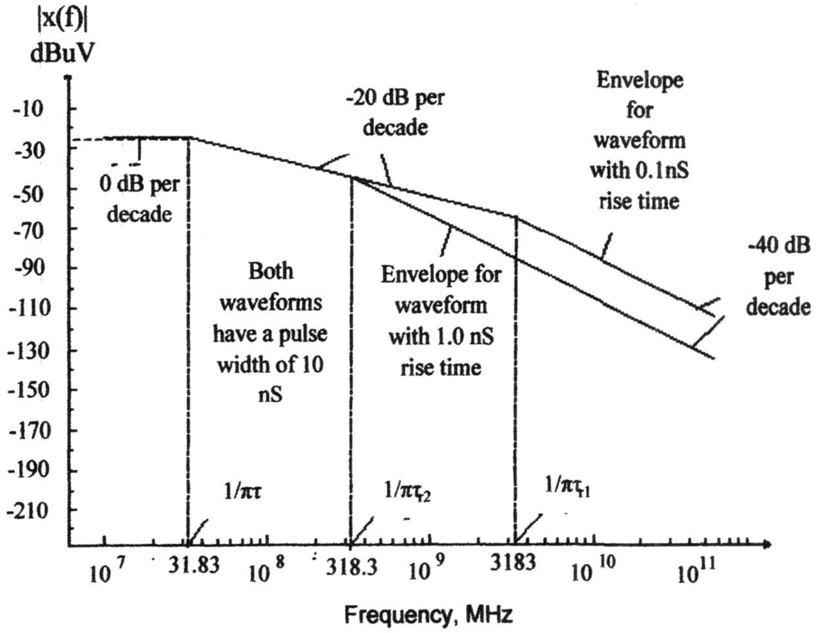

Figure 9.19 **Spectrum Bandwidth for Rise Time (tr) of 1 nS and 0.1 nS**

9.10 ELECTROMAGNETIC ENVIRONMENT

What is unique about today's environment that did not exist (to such a magnitude) with respect to the early automotive systems? The answer is clearly crowded and heavily utilized spectrum of RF communications. Evidence of this can be seen across the landscape with installations that look similar to those shown in the following figures. This shows the multitude of broadcast, cellular and PCS, communication and paging services that are in operation today. In addition to many of these being sources of high field strength levels of emissions, they operate on a multitude of frequencies, further complicating potential EMC issues in automotive systems. Especially in major metropolitan areas, field strength levels may be strong enough to affect the operation of any system that has not been designed or tested for its ability to be immune to these sources of external energy.

Figure 9.20 **Antenna "Farm"**

Figure 9.21 **Cellular Telephone Antenna**

Figure 9.22 **Microwave Relay Tower**

Figure 9.23 **RADAR Installation**

If we look at the EMC tree of disciplines, we can see where radiated immunity falls within the organization of EMC. This portion of the book will discuss aspects of radiated immunity, its characteristics, and typical impact upon automotive systems.

9.11 EMC ISSUE: IMMUNITY TO EXTERNAL FIELDS

One of the most important items to realize is that electronic systems (especially microprocessor and complex systems) operate with various signals and data. Many of these signals consist of operation on various "clock" frequencies that may extend well into the frequency ranges that are being used for RF communication, which may result in systems inadvertently reacting to the external sources of energy instead of the intended signal and/or data that is actually desired. If we look at the RF spectrum, we can understand how this could occur. Figures 9.24 and 9.25 show the many systems and services that use various frequencies, and it is easy to see how the system may react to external sources of energy. Therefore, the goal to address immunity issues is to *understand the compatibility of the electronic systems with the environment.*

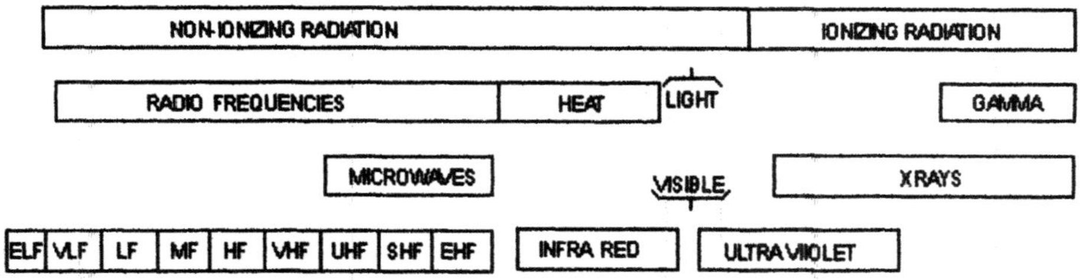

Figure 9.24. **Frequency Spectrum**

Some case studies show how automobile systems have been affected by external sources of RF energy.

9.11.1 Vehicle Anti-Lock Brake System (ABS)

During the early years of ABS, Mercedes-Benz automobiles equipped with ABS had severe braking problems along a certain stretch of the German Autobahn. The brakes were affected by a near-by radio transmitter as drivers applied them on the curved section of highway. The near-term solution was to erect a mesh screen along the roadway to attenuate the EMI. This enabled the brakes to function properly when needed.

While not directly related to automotive systems, this same document also reports on aircraft passenger carry-on devices. Everyone recalls the requests that are heard when on board a commercial aircraft to turn off electronic devices. How real is the threat from those devices? The following are some details on that situation

9.11.2 Aircraft Passenger Carry-On Devices Cases

Passenger carry-on devices provide another group of case histories. They show the increased susceptibility to external EMI sources that modem automated electronic systems aboard aircraft experience. This external EMI is generated by seemingly innocuous electronic devices, which include portable computers, AM-FM "walkman" cassette players, dictaphones, radios, heart monitors, and cellular phones.

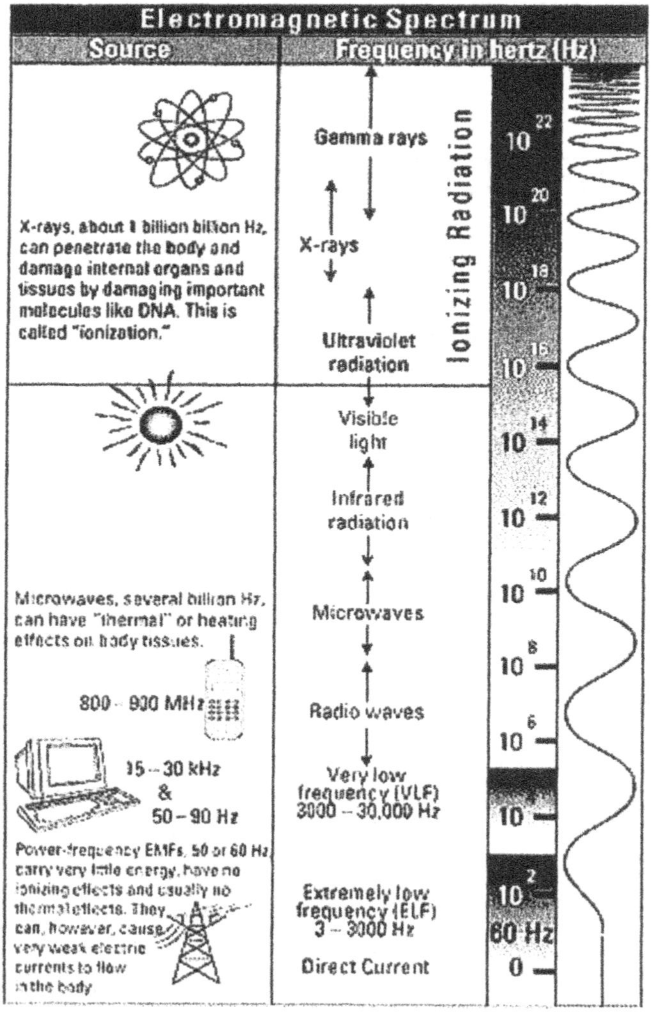

Figure 9.25 **Frequency Allocations**

NASA maintains a database, the FAA Aviation Safety Reporting System (ASRS) and administered by Battelle, which is a compilation of voluntary reports detailing safety problems submitted by pilots or crew members flying a wide variety of commercial and private aircraft. These reports are, for the most part, anonymous with nonspecific aircraft models and unidentified operating companies. At present, the database contains 46,798 full-form reports submitted since January 1, 1986. This author requested an ASRS data base to find all cases that referenced passenger carry-on electronic devices. The results were 56 citations, of which 29 appear definitely to be EMI caused anomalies. Twelve other cases could be EMI related, but additional information is required to make this determination. Table 1 contains a tabulation of the 29 EMI events by affected equipment and suspected cause.

Table 9.1. **Pilot/Crew Reports of EMI Caused by Passenger Carryon Devices**

	Victim Device			
Suspected Cause	Navigation Aids	Communication	VOR	Totals
Cellular Phone	4	1	3	8
Laptop Computer	3	0	2	5
Radio	3	1	0	4
Electronic Game	1	0	2	3
CD Player	0	1	1	2
Tape Player	2	0	0	2
AM/FM Recorder	0	0	1	1
AM/FM Walkman	0	0	1	1
Dictaphone	0	0	1	1
Heart Monitor	0	1	0	2
Television	1	0	0	1
Totals	14	4	11	29

Since automobiles do have ignition and engine control systems, we can also learn about experiences that have been documented with engines and immunity to RF energy. It was reported that a VOA (Voice of America) short-wave broadcast station caused engine failure to occur in the following situation.

9.11.3 F-16 Flight Controls

An F-16 fighter jet crashed in the vicinity of a Voice of America (VOA) radio transmitter because its fly-by-wire flight control system was susceptible to the HIRF (High Intensity Radiated Field) transmitted. Since the F-16 is inherently unstable, the pilot must rely on the flight computer to fly the aircraft. Subsequently, many of the F-16's were modified to prevent this type EMI, caused by inadequate military specifications on that particular electronics system. This F-16 case history was one of the drivers for institution by the Federal Aviation Administration (FAA) of the HIRF certification program.

9.11.4 Blimp Problems

Another VOA transmitter case involved a blimp over Greenville, NC. Flying near the VOA transmitter, the blimp suddenly had double engine failure. The flight crew followed emergency procedures and made a successful unpowered landing. An investigation determined that the failure of the ignition system was extreme EMI. Subsequently, blimps were outfitted with ignition systems protected from HF transmissions.

9.11.5 Boeing 747 Automatic Direction Finder (ADF)

A Boeing employee related this incident. During testing, audio reception on the Boeing 747 communications receivers was unacceptable while the ADF system was in use. Investigation showed that wire-to-wire coupling was the problem because the ADF antenna lead was not separated far enough from other wiring.

9.11.6 Severmorsk Disaster

In mid-May 1984, a Soviet ammunition depot exploded. The cause of the accident, according to the Soviets, was an over-the-horizon radar that had illuminated the depot.

9.11.7 Tornado Fighter Case

Another VOA HIRF case occurred in 1984 near Munich, Germany. A West German Tornado fighter crashed after flying too close to a powerful VOA transmitter.

9.11.8 Libyan Strike

In 1986 during the US air strike on Libya, several missiles failed to strike designated targets and an F-111 fighter crashed. Air Force officials blamed these incidents on EMI caused by U.S. aircraft transmissions interfering with each other.

9.11.9 Antilock Braking System (ABS)

Early ABS systems on both aircraft and automobiles were susceptible to EMI. Accidents occurred when the brakes functioned improperly because EMI disrupted the ABS control system. For aircraft, the initial solution was to provide a manual switch to lock out the ABS function when it was inoperable due to EMI and to use the normal braking system. Later, the solution was to qualify prior to flight the ABS system based on the expected EME. For automobile systems, the solution was to ensure, if EMI occurs, that the ABS system degrade gracefully to normal braking - essentially an automatic version of the aircraft manual switch. Eventually, automobile ABS was qualified by EMI testing prior to procurement.

9.11.10 Fuel System Operation

In the early 1970's, an automobile manufacturer introduced a new fuel injection system, which was sensitive to RF at around 150 MHz. RF could cause a failure in which the electronic fuel injection system released fuel to all of the cylinders simultaneously, instead of releasing fuel to cylinders according to the firing order.

9.11.11 Aircraft

Apparently, operational frequencies of cellular phones, computers, radios, and electronic games are often EMI culprit sources. Many airlines have established rules forbidding or limiting the use of these devices. A special report by the Radio Technical Commission for Aeronautics (RTCA) concluded that a culprit carry-on device has to operate at a frequency that falls within the operating frequency of a particular aircraft system, and that the device probably has to be oriented with its maximum radiation directed out a nearby window for navigation and communication antennas to pick up the emitted radiation. A particularly interesting case from the ASRS search was one where the pilot of a large aircraft actually performed an experiment with a passenger and his laptop computer. The pilot asked the passenger to switch the computer on and off for varying time intervals while he monitored the effect on the aircraft VOR. The pilot reported it was very evident that the computer affected his flight instruments.

One well-known EMI case is not included in the ASRS data base. In February, 1993, a DC-10 autopilot was disrupted during final landing approach by a battery-powered CD player operated by a passenger in first-class. To prevent the aircraft from crashing after suddenly veering off course, the pilot had to manually take control of the aircraft.

9.11.12 Medical Equipment Cases

Modern medical equipment have experienced EMI problems. From 1979 to 1993, the PDA received over 90 reports concerning EMI problems in the field. Silverberg pointed out in his article that users experiencing medical equipment performance degradation may not suspect EMI as a possible cause. Thus, EMI problems are more likely to be under reported to the FDA than other equipment problems.

It is interesting to note that PDA data, like FAA aircraft data, indicate that cellular phones are frequent EMI culprit devices. They have interfered with the operation of incubators, infusion pumps and controllers, dialysis equipment, and defibrillators as well as with aircraft systems. A large hospital in Chicago and a large healthcare center in Indiana have banned the use of cellular phones. These phones are also banned in some European hospitals. Cellular phones that use the new European GSM standard have been reported to produce audible interference in hearing aids up to a distance of 30 m.

Details of several interesting cases concerning medical equipment include

9.11.13 Talking EEG Machine

This case involved EMI that prevented proper testing of surgically implanted probes used in monitoring specific portions of a patient's brain activity. With probes in direct brain contact, the potential between any two points is measured on an EEG machine. The EEG provides critical feed-back to the surgeon during surgery. This particular EMI manifested itself on the analog plotting needles of the EEG machine as a modulated signal easily recognized as speech - hence a talking EEG machine! The EMI-caused noise was so severe that it completely masked the EEG signals and made the machine alarmingly ineffectual during surgery. The signal was from a local AM radio station, and the noise during surgery was from common impedance coupling between the EEG machine and the operating table. Bonding the EEG with the operating table eliminated the EMI and restored the critical brain monitoring function.

9.11.14 Ambulance Heart Monitor/Defibrillator

Susceptibility of medical equipment to conducted or radiated emission is a concern. In this case, a 93-year-old heart attack victim was being taken to the hospital and the medical technician had attached a monitor/defibrillator to the patient. Because the machine shut down every time the technicians turned on the radio transmitter to request medical advice, the patient died. An investigation showed that the monitor/defibrillator was exposed to exceptionally high radiated emissions because the ambulance roof had been changed from metal to fiberglass and fitted with a long-range radio antenna. Reduced shielding combined with the strong radiated radio signal resulted in EMI to the vital machine.

9.11.15 Runaway Wheelchairs

Wheelchairs came under the scrutiny of the FDA because of reported erratic, unintentional powered-wheelchair movements. These movements included sudden starts that caused wheelchairs to drive off curbs or piers when police, fire, or CB transmitters were activated near the chairs. Although no fatal injuries have been reported, FDA has ordered manufacturers of motorized wheelchairs to shield them from EMI and to educate users on the potential EMI hazards.

9.12 INEXPENSIVE SHIELDING METHODS

What could be one way to eliminate the effect of external energy upon electronic devices? There are some methods that can be used, unfortunately, many of these methods would require additional time or re-design of the product, which in many cases is not possible in the tight schedule of automotive system development. One possible method may be the use of a shield around the device to reduce the amount of energy that is received by the device itself. In this case, if we have sources of RF transmitters, these are typically in terms of v/m, and this would mean that we would want to provide shielding against electric field energy. A very effective shield can be made from thin pieces of aluminum, sometimes even as thin an aluminum foil that is used in food storage.

The ability of this type of corrective action to provide protection can be seen in a simple demonstration that one of the authors uses in discussions on immunity. The parts of that demonstration appear in Figure 9.26. The parts are a CD player (to represent a complex piece of electronic equipment), external speakers (to identify when the player is in operational mode), and a VHF/UHF transmitter. The test process is as follows. The CD is playing a disc, and the transmitter is set to a frequency of approx 150 MHZ. It is then brought near the player, and the energy at about six to nine inches away causes the player to malfunction. This same approach is used on the frequency of about 450 MHz, and the player also stops working.

A shield is then constructed by wrapping the foil around the player. This is also shown in the Figure 9.27. The shield is effective at 150 MHz, and is not effective at 450 Mhz. The reasons for this will be discussed later and indicate that each EMC solution needs to be understood as to it's ability to address the different frequencies that may be involved.

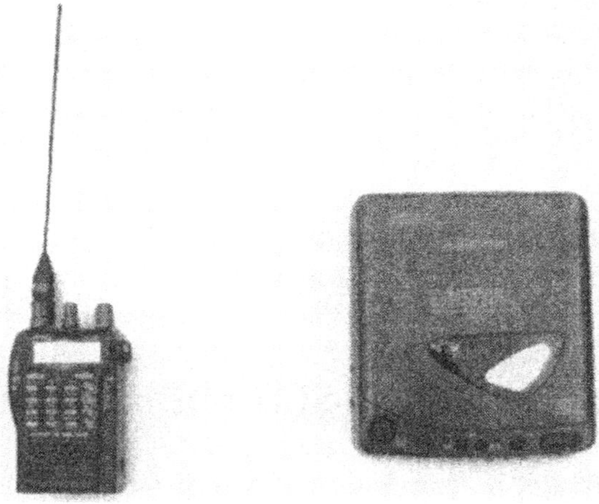

Figure 9.26. **Transmitter Interferes With CD Player**

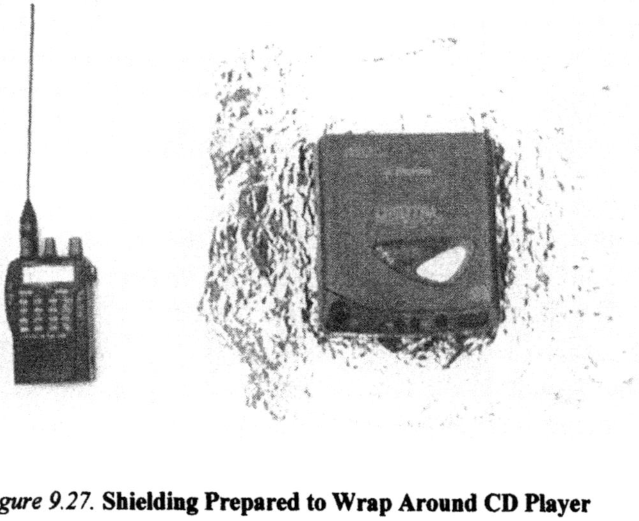

Figure 9.27. **Shielding Prepared to Wrap Around CD Player**

Figure 9.28. **CD Player Enclosed in Shielding**

Another case study is documented on a vehicle with a two-way radio that seemed to have problems with transmission operation when being driven. The vehicle was returned to the dealer several times, and each time the dealer was unable to completely identify the cause of the problem. They finally decided that the transmission needed to be replaced. (An expensive fix!) During the time that the plan to replace the transmission was agreed upon, the vehicle was taken to a two-way radio shop for removal of its radio. Over the next few days, the owner of the vehicle noticed that the vehicle performed without any problems. This then led the owner to conclude that the radio was affecting the transmission. The radio was re-installed, and the problem reappeared. While the vehicle was being examined by the dealer, it was noticed that the engine computer was in a plastic-housed box. A computer housed in a metal box from another vehicle was installed, the two-way radio was utilized, and the vehicle performed without problems. The end result was that perhaps in the interest of cost-savings, the plastic box was deemed to be appropriate for all the performance requirements of the computer – except EMC!

9.13 EMC DESIGN FOR IMMUNITY

There are design-related approaches that can be used to increase the immunity characteristics of automotive electronic components and systems. Many of these approaches will have minimal impact upon component functionality and performance; however, note that, for maximum efficiency and minimal cost, they need to be included in the design stage. The advantage of some of these is that they may be included at later stages in the product development process, if EMI issues are discovered.

These design features are summarized as follows:

- Add series inductance to sensitive I/O lines

  Reason: adding inductance may produce enough attenuation of the RF energy to prevent immunity problems from occurring, and if a minimal amount is used, it should not have an impact upon component functionality.
- Add parallel capacitance to shunt RF away from sensitive lines.

  Reason: Adding parallel capacitance will develop a lower impedance path for the energy to take rather than disrupting sensitive electronic devices. Many times these can be incorporated, since they are in parallel with the lines.
- Buffer or isolate circuits from each other.

  Reason: This will help to eliminate any current loop conditions that might exist due to the RF energy into the system.
- A design criteria item (although difficult after the device has been designed) is to utilize only the minimum gain-bandwidth of the system that is needed to process the data.

  Reason: By having high gain-bandwidth, this may allows noise (which is typically high frequency) to easily enter the system and cause problems.
- Add localized shielding to the circuit.

  Reason: eliminate the energy from entering the circuit, although this can be a very expensive and not totally effective fix (as was seen in the CD player discussions). The shield may fall off or not be installed, or may be damaged.

9.13.1 Component Selection

One of the basic building blocks for any electrical design is selection of the components. Selection of components for EMC applications is equally as important as selection for performance. Except for wideband video and circuits employing oscillators, analog circuits are generally much quieter than digital circuits. Because digital circuits are noisier, this section emphasizes the selection of digital components for suppression of EMI. The most important issue in selecting digital components for low-noise characteristics is rate of change of energy. The noise voltage induced into a victim circuit from a noise source circuit is:

$$V = -M\, dI/dt$$

where M is the mutual inductance between the two circuits and the coupling is magnetic in nature.

Or:

$$V = C\, dV/dt$$

where C is the capacitance between the two circuits, and coupling is electric in nature. Mutual inductance, M, depends on current loop areas of source and victim, orientation, separation distance, and the heights of the circuits above ground. Source and victim current loops are analogous to the primary and secondary windings of a transformer (fig.9.29). Capacitance, C, depends on the distance between conductors, associated effective areas, and Z, the impedance to ground of the victim circuit. The source and victim conductors act as a parallel plate capacitor (fig. 9.30).

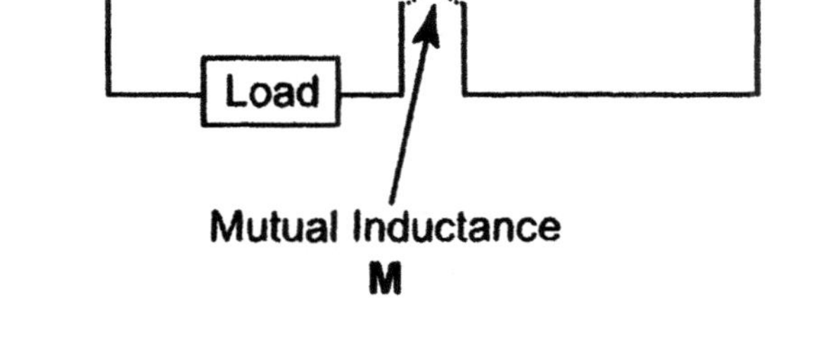

Figure 9.29. **Noise Coupling Via Mutual Inductance**

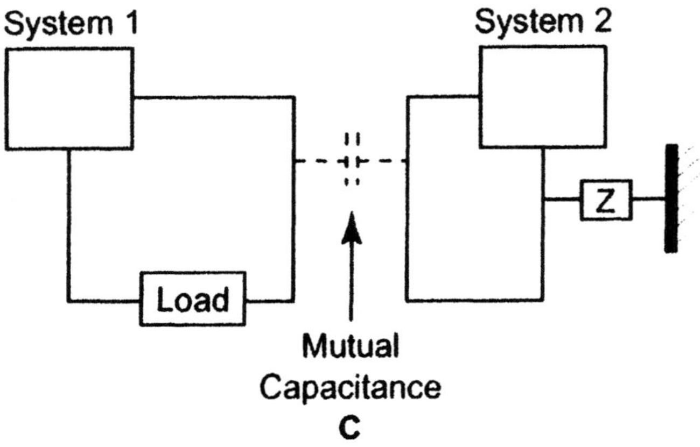

Figure 9.30. **Noise Coupling Via Electric Induction**

9.13.2 Logic Families and d*V*/*dt*

Table 9.1 shows various digital family rise times and voltage rates of change (d*V*/*dt*). The faster the rise time and the higher the voltage swing, the larger the d*V*/*dt*. Using the slowest rise time to achieve the desired function can lower the amount of noise coupling. Another reason for using slower rise time is to limit the higher frequency harmonics of the digital signal. Because the circuit traces on printed circuit boards (PCBs) can act as antennas and radiate noise at higher frequencies, limiting the unnecessary harmonics in a digital signal prevents radiation of these higher frequency harmonics. The next section addresses the transformation of time-domain signals into the frequency domain and how slower transition times and lower repetition rates lower and/or eliminate higher frequency harmonics.

Table 9.2. **Rise time and voltage rate of change for various logic families.**

Logic Family	Rise Time (ns)	Voltage Swing (V)	d*V*/*dt* (V/ns)
CMOS 5 V	100	5	0.5
CMOS 12 V	25	12	0.48
CMOS 15 V	50	15	0.30
HCMOS	10	5	0.5
TTL	10	3	0.3
ECL 10k	2	0.8	0.4
ECL 100k	0.75	0.8	1.1

9.13.3 Logic Families and *dI/dt*

As a result of stacking the output stage of the logic circuit in the chip (Figure 9.31), when the logic is switched, the transistors typically turn off slower than they turn on and draw large amounts of transient current from Vcc during the transition. This induces transients on the Vcc trace and ground. Notice that the output stage of the TTL circuit contains a current limiting resistor. The CMOS circuit has no current limiting resistor and, consequently, draws larger currents *(dI/dt* sometimes as high as 5,000 A/s) than TTL. One way to limit these surges is through the use of decoupling capacitors. The decoupling capacitor, which will supply the necessary instantaneous currents while the chip is switching, is a capacitor connected between Vcc and ground (Figure 9.32). It is important to remember to make the capacitor leads as short as possible to reduce parasitic inductance and to mount the capacitor close to the decoupled chip to reduce loop area.

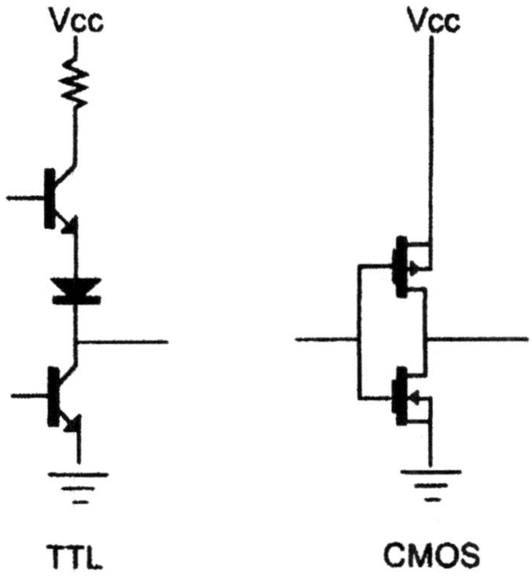

Figure 9.31. **Logic Output Drivers**

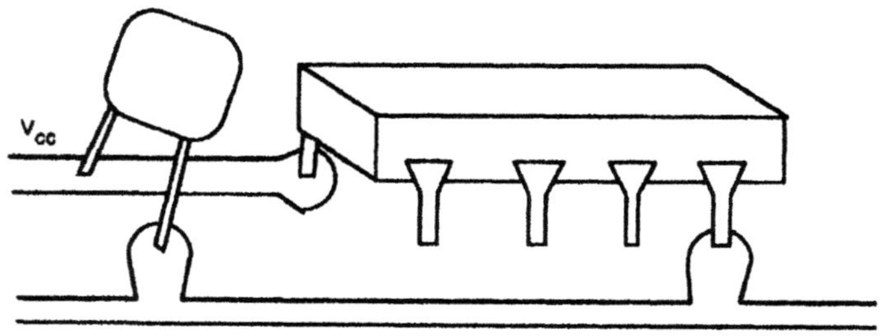

Figure 9.32 **Capacitor Connected from Vcc to Ground**

9.14 IMMUNITY THRESHOLD

Use of a hand-held transmitter can be a good way to determine the immunity characteristics of electronic devices, and it turns out that we can calculate the approximate field strength of the transmitter, which we can take advantage of to determine both the approximate absolute values of the device's immunity and relative amounts of improvements that are needed or made.

The approximate values of field strength from a hand-held transmitter can be calculated from the following formulas

E= 5.5 x SQRT (Power) / Distance from the transmitter.
E is v/m
Power is Watts
Distance is meters

It is important to understand that, for most conditions, this will mean that the device under test will be in the near-field environment. We also should remember that the absolute values would vary as a result of the type of antenna used on the transmitter. For example, the flexible antenna typically used on hand-helds has a loss of anywhere from 3-10 dB. This would mean that the field strength would be reduced.

In most situations, consumer devices will exhibit immunity sensitivities at about field strength of 3 – 10 v/m. This is well within the capability of hand-held transmitters to produce. This means if we take an immunity threshold level of 5 V/m, we can approximate the distance that may cause problems as $D = \sqrt{P}$ (based on $5 = (5.5\sqrt{P})/D$).

The process to determine the effectiveness of various fixes or corrective actions would consist of finding out what the threshold of failure is, in terms of distance from the transmitter, and perhaps estimating the field strength. Then the changes would be made, and the same process would be reported to determine the amount of improvement made. From previous chapters, we know how we can convert the absolute improvements in terms of dB. This process can work well to test various wiring and harnessing connected to the device under test, as well as to determine the shielding effectiveness (to be discussed later in the text).

It also turns out that many immunity issues manifest themselves at about 150 MHz, which is a convenient frequency that many hand-held radios produce, and do pose a real-world threat.

9.15 AUTO INDUSTRY "BEST PRACTICES"

As a result of the interaction of wiring and cabling on vehicles, there has been a list of "best practices" that has been developed to assist in the vehicle level engineering. As with all best practices, the key is to understand that not all practices apply to all situations all of the time. It is important to understand the rationale behind each practice, to see if it makes sense to apply the particular item for a specific application. The following list is not all inclusive of every wiring best practice – it is a subset of some key ones to help the reader understand what their own organization might have in place for similar types of issues.

Table 9.3 shows some best practices for wiring.

Table 9.3 **Wiring Best Practices**

Practice	**Rationale**
Route wiring away from ignition, System secondary wiring and spark plugs	Noise may couple via E-field coupling
Don't bundle antenna, speaker, and radio power wiring with vehicle wiring	Noise may couple via H-field

9.16 IGNITION SYSTEMS

9.16.1 Spark Plugs

While the insertion of a resistor-suppressor in series with the spark plug is a sound method for the suppression of extraneous oscillations, there has been considerable apprehension as to the possibility of impairing the efficiency of the ignition system because of the reduction of the peak energy of the capacitive component of the spark discharge. Exhaustive tests have shown that resistor-suppressors in values up to 150,000 ohms, many times greater than used in suppression systems, have practically no effect on torque, fuel economy, or horsepower output. Figure 9.33 shows the ratio of fuel consumption to brake horsepower for three spark plugs, the first with no resistor-suppressor, the second with a 10,000-ohm resistor-suppressor, and the third with a 20,000-ohm resistor-suppressor.

Another misconception concerning resistor-suppressors is that they intensify the problems of cold starting. Results of carefully controlled tests at the U.S. Signal Corps Engineering Laboratories have demonstrated that at –30 degrees F and colder suppressors are no deterrent to cold starting and, in fact, in some cases permit almost instant starting when the identical motor without suppressors would start only after several minutes of cranking. Apparently, the inductive component in the ignition spark discharge is more likely to affect the combustion of an extremely cold mixture than the capacitive component; with the inductive component appearing as a continuous current flow with a resistor-suppressor installed, the greater time at less total spark energy appears to be more effective in igniting cold gas than the higher heat energy in a more instantaneous spark. The results of these laboratory tests are illustrated in Figure 9.34.

An unexpected advantage found with suppressor-equipped spark plugs is that the gap growth is considerably less than with the same type plug without a suppressor. Tests on several different engines running continuously at wide open throttle at various speeds reveal that the rate of gap growth in spark plugs with 10,000-ohm suppressors is about one-half that experienced with standard plugs. The results of these tests are illustrated in Figure 9.35.

It is desirable for all combat vehicles as well as engine generators and other engine-driven equipment to be furnished with integrally suppressed spark plugs, shown in Figure 9.36, in which resistor-suppressor elements are built into the plugs. The advantages of applying the suppressors in this manner are:

Integrally suppressed sparkplugs provide a higher degree of suppression effectiveness because of the absence of any length of exposed high-tension cable on the "hot" side of the suppressor, between suppressor and gap, which is required when external suppressors are used and may radiate considerable interference.

The stock and supply problems are reduced since no separate suppressors are required.

The suppressors cannot be inadvertently or deliberately omitted by maintenance personnel.

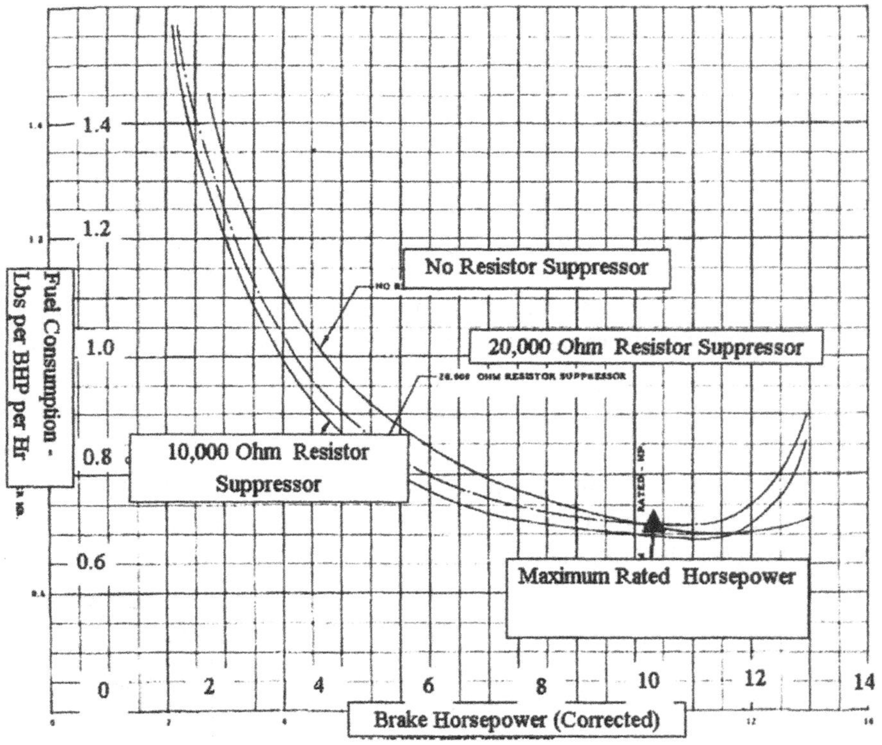

Figure 9.33. **Effect of Resistor Suppressor on Engine Performance**

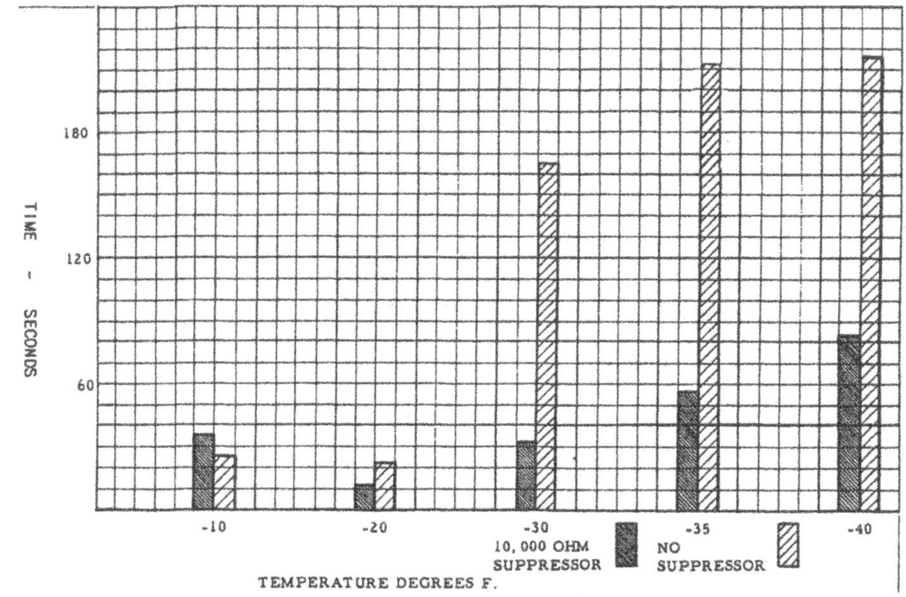

Figure 9.34. **Average Starting Time; Suppressors at Spark Plugs vs no Suppressors**

While integrally suppressed spark plugs are preferred, in applications where the spark plugs are not so suppressed, the application of an external resistor-suppressor to a plug may solve a radio-interference problem. The installation of an external resistance suppressor is shown in Figure 9.37

Despite the success of integrally suppressed spark plugs in reducing ignition interference from the inductive component of the spark discharge, they do not entirely eliminate it. In addition, the capacitive component is responsible for steep wave transients in the secondary circuit even though the peak energy is reduced. These two factors necessitate complete shielding of the high-tension circuit in the ignition system. Self-shielded plugs are preferred because of the reduced possibility of leakage of radio-interference currents.

The use of unshielded spark plugs is to be avoided for all equipment procured by the Department of the Army except for administrative-type vehicles. Shielded sparkplugs are readily available for almost all current ignition installations.

Figure 9.35. **Spark Plugs, Standard and Integrally Suppressed, Showing Increased Gap Growth in Standard Spark Plug. (From top, (A) Standard Type, 222 Hours. (B) Integrally Suppressed Brand "A", 529 Hours. (C & D) Integrally Suppressed Brand "C", 529 Hours**

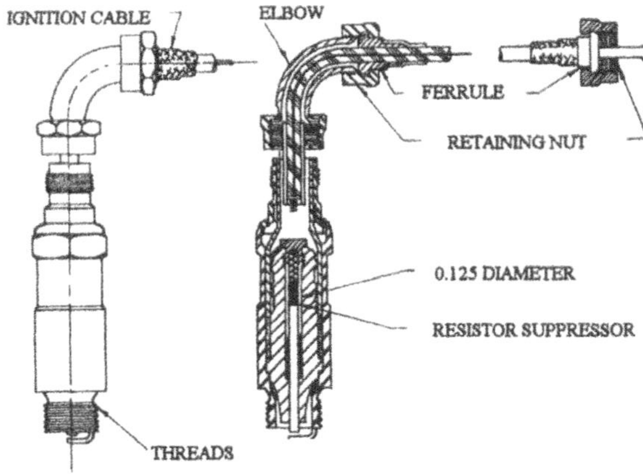

Figure 9.36. **Integrally Suppressed Spark Plug**

9.16.2 Distributors

Distributors are essentially switching devices in which a rotor distributes the energy received from the coil to the various spark plugs.

The distributor has two functions which can be severe sources of radio interference: the switching function which causes a transient or variable current state, and the arcing which occurs between the rotor and the electrode during the switching process.

Recommended capacitor installation. is that the capacitor be installed externally. The lead from the coil terminal to the capacitor should be kept as short as possible.

The lead from the coil to the distributor is in the high-tension circuit and must be shielded. The installation of resistor-suppressors at the distributor and at the spark plug makes braided shielding adequate for this lead.

As was discussed previously regarding distributors, the preferable installation is to incorporate the coil and distributor in a single housing. A feed-through capacitor is used in the battery supply lead. The lead between the coil and distributor is entirely enclosed within the shield, making the use of braid unnecessary. This system incorporates the recommended practices of interference suppression and, in addition, minimizes interconnecting wiring and the possibility of leakage.

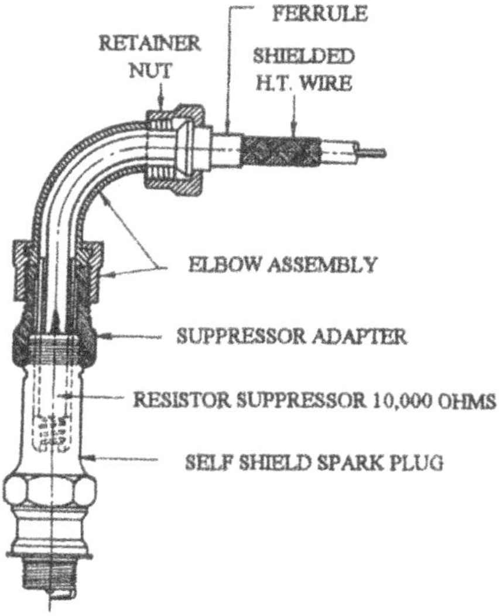

Figure 9.37. **External Resistor Suppressor Installed at a Spark Plug**

9.16.3 Ignition Harnesses

Ignition harnesses are not a source of interference in themselves; however, they can serve as means for radiation of interference from other sources. Steep transients are present in the interconnecting wiring in the ignition system as a result of the action of the spark plugs, distributor, and breaker points; and even though these components may be well shielded, the interconnecting wiring can radiate the interference. As a result, as much care must be taken with the suppression of interference emanating from the interconnecting wiring as with the components themselves.

The ignition harness can be divided into two categories; the high-tension and low-tension wiring. High-tension wiring includes the lead from the coil to the distributor and from the distributor leads to the sparkplugs. The low-tension wiring includes the battery supply and the breaker-points lead to the coil. The high-tension wiring in the secondary circuit is more important because of the higher voltages and steeper transients involved. These transients can be reduced by the installation of resistor-suppressors in the secondary circuit; however, some degree of shielding is necessary for the high-tension wiring to prevent the radiation of interference.

Shielding for high-tension ignition wiring may vary in construction from low-percent coverage, loosely woven wire braid with relatively poor shielding effectiveness, through double and triple layers of wire braid, to completely solid wall conduit, where shielding effectiveness may be increased to any desired degree by increasing wall thickness. The degree of shielding depends on the amount of interference that necessitates suppression.

Flexible shielding conduit gives high percent coverage and is effective in preventing the radiation of interference from ignition wiring. It is made of strip metal formed either into spiral bellows or into some other kind of spiral that allows interlocking of adjacent strips. It may either be soldered at the seams or allowed to provide sliding action between turns. For more effective shielding, it may be covered with one or more layers of woven metal braid.

Flexible shielding conduit has disadvantages which necessitate holding its use to a minimum. Some of these are:
It is expensive.
It requires the use of considerable quantities of strategic copper, brass, and/or bronze.
Machine tool manufacturing facilities are tied up in its manufacture.

The addition of resistor-suppressors in the secondary circuit, which reduces the steep transients, makes it possible to substitute tinned-copper braid shielding for flexible conduit shielding, which is less expensive, uses less strategic materials, and requires less care in production and maintenance. Figure 9.38 shows the construction of tinned-copper braid shielding for a high-tension ignition cable

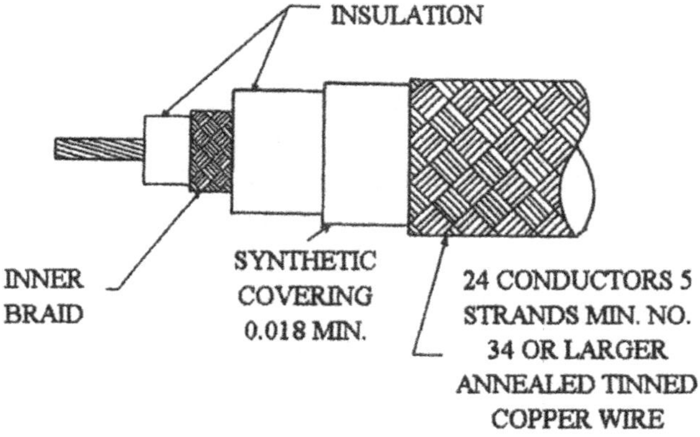

Figure 9.38. **Shielded High Voltage Ignition Cable**

Chapter 10

EMC Regulation of Automotive Systems

10.1 INTRODUCTION

This chapter can be considered a "good news / bad news" chapter! The good news is that for the most part, automotive systems are not directly regulated by governmental legislation, unlike most other consumer electronic devices. The bad news is that the industry (especially in the United States) has a responsibility to "self-police" itself in order to maintain it's relatively regulation-free status! Even in those countries that governmental regulations do apply, this is the exception rather than the rule and typically the regulations are at a value that if only met, would be at a severe customer dissatisfaction.

10.2 RADIATED EMISSIONS REQUIREMENTS

It has turned out that to date, the auto industry is exempt from many radiated emissions requirements that are required for other consumer electronic products. Where it does exist, we need to comprehend what the requirements are as there are some automotive systems and components that are subject to legislated requirements. In addition to legislated (or governmental) requirements, customer-based requirements are established by OEMs (original equipment manufacturers). Almost every OEM has requirements in place for radiated emissions from components or systems on their vehicles. There are also governmental requirements that apply to some systems on vehicles, which are covered by standard, or consumer-driven radiated emissions requirements.

10.3 GOVERNMENTAL REQUIREMENTS

The governmental requirements vary from country to country regarding frequencies and emissions levels. In the United States, the Federal Communications Commission (FCC) has the authority to establish and enforce radiated emissions requirements from electronic devices. These are contained in the Commission's rules from the Office of Engineering and Technology and are documented in the "Title 47" portion of the Federal Rules and Regulations. Within "Title 47" is "Part 15", which covers radio frequency emissions. The FCC defines radio frequency as emissions that take place at a frequency between 9 kHz and 3000 GHz. Part 15 itself is intended to control emissions from devices that may produce harmful interference due to both unintentional (classified as incidental) and intentional radiators. The sections of Part 15 that apply to vehicles are shown in Figure 10.1.

As can be seen, incidental emitters on vehicles are generally exempt from Part 15. This means devices such as engine controllers and portions of other systems. The various intentional emitters, such as low power transmitter are subject to Part 15, and are called "non-exempt" devices. A unique requirement for electric vehicles is Part 18, since the system becomes part of the power grid when it is connected to a charging source.

10.4 FCC PART 15

The decade of the 1970's saw tremendous advances in digital electronics. In 1979 the FCC established rules for digital device radiated emissions for "clock speeds greater than 9 kHz". For any digital devices that are marketed today (9 kHz is several orders of magnitude slower than today's devices), all such devices marketed must meet the FCC requirements for both conducted emissions and radiated emissions. Devices are classified according to their intended operating environment as either Class A (commercial location) or Class B (residential location). The requirements for radiated emissions are more stringent for Class B devices than they are for Class A devices. A good example of this is that a computer used or installed in the home should not interfere with a radio or television reception in a neighbor's home.

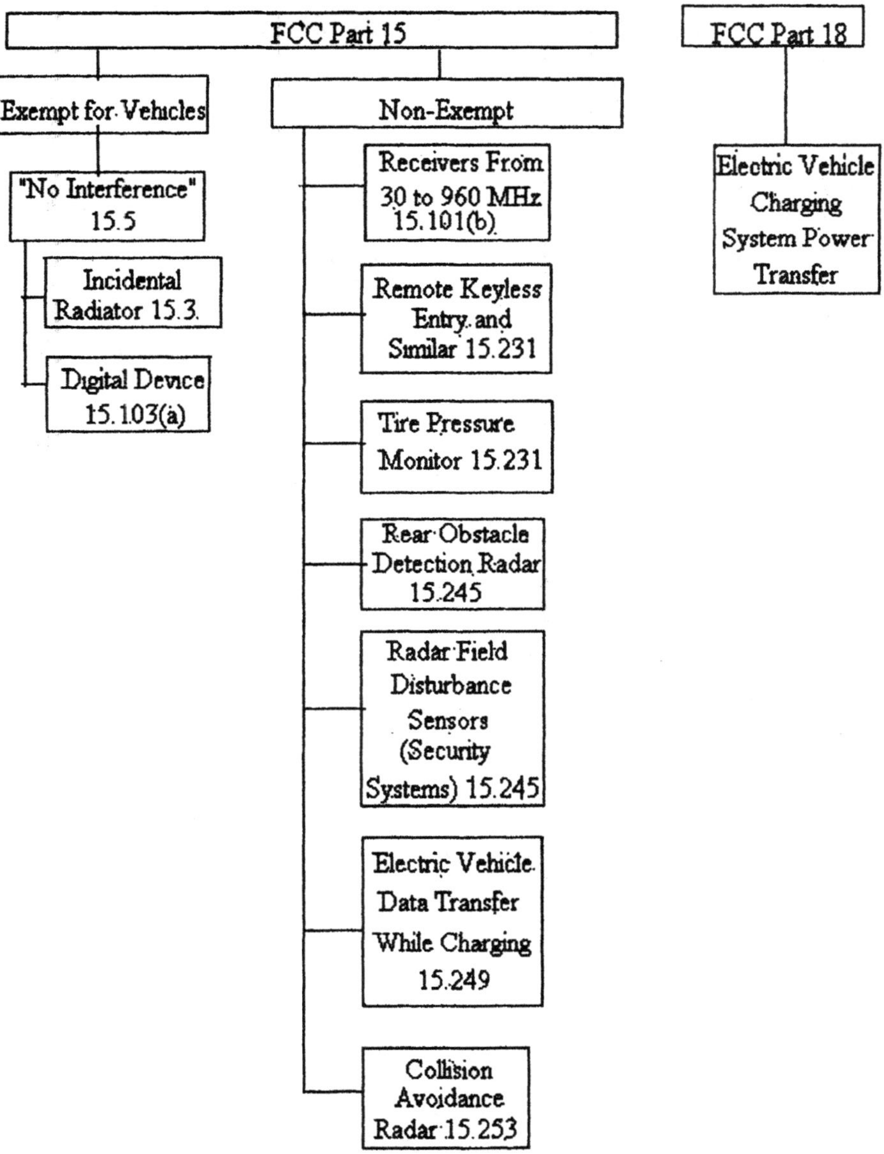

Figure 10.1. **FCC Title 47 Parts Application to Vehicles**

In a residential environment, it is likely that computers and radio receivers will be located closer to one another than in a commercial location; therefore the more stringent requirements. The requirements for radiated emissions cover frequencies from 30 MHz to 40 GHz and are shown in Figure 10.2.

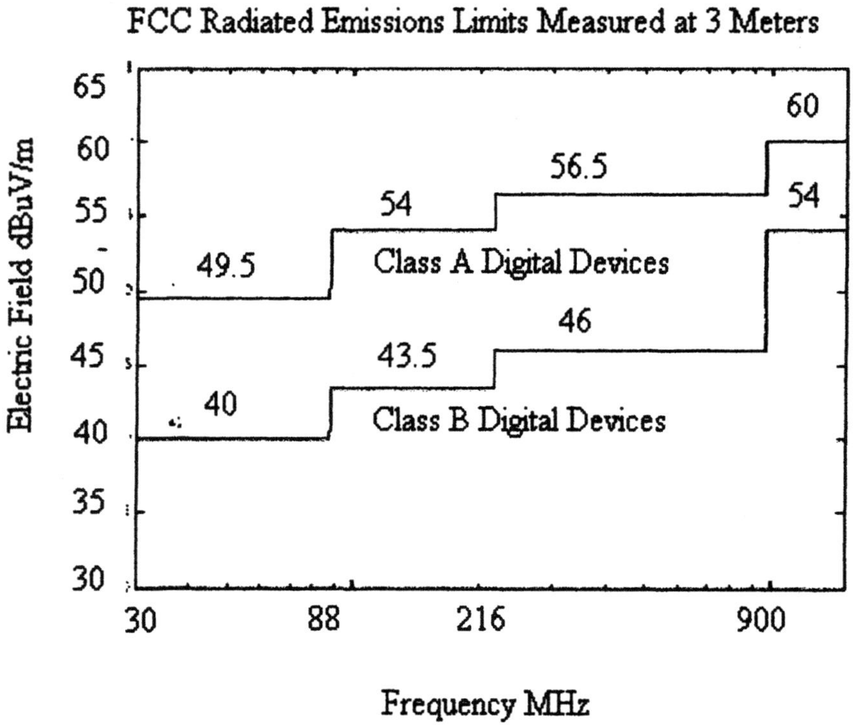

Figure 10.2. **FCC Part 15 Class "A" and Class "B" Radiated Emissions Requirements Normalised to 3 Meters**

In addition to radiated emissions, requirements are also given for conducted emissions. These are specified in the range 450 kHz to 30 MHz. The FCC describes the test method to be used when measuring conducted emissions, or radiated emissions. These test procedures can be seen in a review of FCC Part 15.

Another classification within the FCC regulations has to do with the classification of devices as “unintentional,” or "intentional"

The introduction to Part 15, Subpart A, is “general” and discusses the intent of Part 15 and what types of devices are covered. Subpart B refers to the "unintentional" radiators of energy. This means that the radiation is a by-product of the operation of the device rather than the primary purpose of the device. For example, emissions from a computer are unintentional as the purpose of the computer is to be a computer and not a transmitter. Similarly, a vehicle ignition system would be an unintentional radiator, since the radiation is a consequence of the spark ignition system. Sub-part C covers "intentional" radiators as sources of radio frequency energy. This would include such devices as cellular phones, remote keyless entry transmitters, wireless tire pressure transmitters, and other similar devices, because the intention of these devices is to emit radio frequency energy to carry information. With respect to unintentional radiators, it is significant that over the past few years, the increases in electronic devices' clock speeds have made requirements for controlling higher frequency harmonics important.

Even though Part 15 has many exemptions for automotive systems it is important for us to understand the basic intent of Part 15 and how it relates to the automotive industry. Within FCC Part 15, Paragraph 15.103 defines "exempted devices" and states the following:

[Code of Federal Regulations]
[Title 47, Volume 1]
[Revised as of October 1, 2001]
From the U.S. Government Printing Office via GPO Access
[CITE: 47CFR15.103]

TITLE 47—TELECOMMUNICATION COMMISSION

" Subpart B--Unintentional Radiators

Sec. 15.103 Exempted Devices.

The following devices are subject only to the general conditions of operation in Secs. 15.5 and 15.29 and are exempt from the specific technical standards and other requirements contained in this part. The operator of the exempted device shall be required to stop operating the device upon a finding by the Commission or its representative that the device is causing harmful interference. Operation shall not resume until the condition causing the harmful interference has been corrected. Although not mandatory, it is strongly recommended that the manufacturer of an exempted device endeavor to have the device meet the specific technical standards in this part.

(a) A digital device utilized exclusively in any transportation vehicle including motor vehicles and aircraft.

(b) A digital device used exclusively as an electronic control or power system utilized by a public utility or in an industrial plant. The term public utility includes equipment only to the extent that it is in a dedicated building or large room owned or leased by the utility and does not extend to equipment installed in a subscriber's facility.

(c) A digital device used exclusively as industrial, commercial, or medical test equipment.

(d) A digital device utilized exclusively in an appliance, such as a microwave oven, dishwasher, clothes dryer, air conditioner (central or window), etc.

(e) Specialized medical digital devices (generally used at the direction of or under the supervision of a licensed health care practitioner) whether used in a patient's home or a health care facility. Non-specialized medical devices, i.e., devices marketed through retail channels for use by the general public, are not exempted. This exemption also does not apply to digital devices used for record keeping or any purpose not directly connected with medical treatment.

(f) Digital devices that have a power consumption not exceeding 6 nW.

(g) Joystick controllers or similar devices, such as a mouse, used with digital devices but which contain only non-digital circuitry or a simple circuit to convert the signal to the format required (e.g., an integrated circuit for analog to digital conversion) are viewed as passive add-on devices, not themselves directly subject to the technical standards or the equipment authorization requirements.

(h) Digital devices in which both the highest frequency generated and the highest frequency used are less than 1.705 MHz and which do not operate from the AC power lines or contain provisions for operation while connected to the AC power lines. Digital devices that include, or make provision for the use of, battery eliminators, AC adaptors or battery chargers which permit operation while charging or that connect to the AC power lines indirectly, obtaining their power through another device which is connected to the AC power lines, do not fall under this exemption.

(i) Responsible parties should note that equipment containing more than one device is not exempt from the technical standards in this part unless all of the devices in the equipment meet the criteria for exemption. If only one of the included devices qualifies for exemption, the remainder of the equipment must comply with any applicable regulations. If a device performs more than one function and all of those functions do not meet the criteria for exemption, the device does not qualify for inclusion under the exemptions.

Thus it can be seen that digital devices used exclusively on transportation vehicles, such as engine control modules, and having no application outside the vehicle, would be exempt from the measurements and documentation required by FCC Part 15. Other devices that may be used on a motor vehicle, however, that intentionally radiate RF, such as remote keyless entry systems, security systems, tire pressure monitors or similar devices would be subject to the measurements and documentation described in FCC Part 15 since they are intentional emitters of RF energy. (There are other industries that also have exemptions for FCC Part 15. See the excerpt from Section 15.103).

While it may appear to be good news for the automotive industry that some devices are exempt from Part 15, this is not the case, since more stringent self-imposed requirements are needed to meet customer expectations. If manufacturers were to implement digital devices that emitted radiation at levels that meet the FCC limit, vehicle emissions would result in customer dissatisfaction with entertainment, cellular telephone and other electronics. It is also ironic that many EMC classes, seminars, and design techniques address how to design to meet FCC Part 15. While appearing to be helpful to the automotive industry, in fact, the automotive environment is one of the most demanding environments, and automotive devices must in practice be tens of dB lower than the FCC Part 15 requirements! Just meeting Part 15 for an automotive system is not good enough for most applications!

With regard to specific interpretation of FCC Part 15, one of the authors had an opportunity to personally discuss this issue with FCC personnel. The contact at the FCC explained that the best way to identify the things that are subject to Part 15 is to list the items that are exceptions and then identify those items that remain. The FCC specifically indicated that since the ignition system does not intentionally generate RF energy and it does emit RF energy as a by-product of its operation, the ignition system is an incidental radiator. While there are no specific standards, incidental radiators are subject to the non-interference requirement, which means that if your device causes harmful interference, you may be required to cease operation of the device. Again, the digital devices used exclusively in transportation vehicles are exempt from the technical standards themselves and are also subject to the non-interference requirements. The contact at the FCC also stated that the other devices in the automobile are subject to Part 15 standards. The bottom line is that if a device is an unintentional radiator, and used on a vehicle, the device is exempt from Part 15. If the device is an intentional radiator, it is subject to regulation by Part 15.

As a side matter, the contact at the FCC explained that the Commission's written requirements are expressed in the unit's microvolts per meter and that the graphical requirements are converted to dB microvolts per meter (as shown in the conversion equations in another chapter of this text).

The FCC Part 15 regulations have different requirements with regard to the upper range of the frequency emissions that must be measured. There are five different frequency ranges for emissions measurements that may be imposed. These upper ranges are from 30 MHz for very low frequency sources (about 1 MHz) to the fifth harmonic or 40 GHz (whichever is lower) for high frequency operated devices. The details of each of these ranges can be seen in Table 1.

Table 10.1. **Frequency Range Over Which Radiated Emissions Must Be Measured**

Clock Frequency	Upper Limit of Frequency Range Over Which Measurement Must be Made
<1.705 MHz	<30 MHz
1.705–108 MHz	1000 MHz
108–500 MHz	2000 MHz
500–1000 MHz	5000 MHz
>1000 MHz	lower of: 5th harmonic or 40 GHz

Another important thing to understand about the difference in the requirements for Part 15 is the different measurement distance specified for Class A and Class B. Recall that the difference between Class A and Class B devices are those are intended to operate in a commercial environment or a residential environment, respectively. It is reasonable that the emissions from those devices are also measured at different distances, to represent the operation of the device in different environments. This is why the emissions from Class A devices are specified as to take place at a range of 10 meters from the device, and the measurement of Class B devices is specified to take place at a range of 3 m. This makes sense as Class A devices are utilized in commercial locations, and commercial locations may have a higher degree of separation between the digital device and a receiver that might be affected by the emissions. Class B devices are used in a residential environment, so might be located rather close to a victim receiver, such as a computer being located in one room, close to a radio receiver in an adjacent room.

Look at the limits in Figure 1 for the emissions for each class of devices at the lower frequency end (approximate 30 MHz). If we normalize the measurement distances to 3 m, Class B devices are permitted 9 dB less emission then are Class A devices. This normalization is accomplished due to the fact that the field strength (in the far field) falls off as a reciprocal of the distance from the source. This is represented as the following relationship:

New Field Strength / Old Distance = Original Field Strength / New Distance

For example,

30 V/m at 10 meters distance = 15 V/m at 20 meters distance

As we move higher in frequency to approximately 1 GHz (actually 960 MHz) the difference is about 6 dB. If you think back to the work we've done in other chapters, you'll realize that the 6 dB difference is actually a doubling of electric field strength between the Class A and Class B requirements.

To summarize, there are different requirements in the world corresponding to different national jurisdictions. In United States the FCC has responsibility and authority to control radiated emissions. In some other countries (Canada, etc.) and in some other regions (Europe), different regulations and standards are in force. It is important to realize that there is no uniform worldwide standard with regard to radiated emissions for products that are to be marketed in different regions of the world or n different countries. It is the manufacturer's duty (often delegated to the vehicle system supplier) to comply with specific local requirements. In addition to radiated emissions there may also be requirements for conducted emissions.

It is interesting to note that Part 15 is not stringent enough for the demands of automotive vehicles. Specifically, low levels of radiated emissions are required in automotive systems and components, as compared to many consumer devices, due to the proximity to each other.

Many of the automotive manufacturer's internal specifications for component level radiated emissions are approximately 10- 30 dB more stringent then those that would be allowed under typical national or country regulations, or specifically FCC Part 15! This is because the regulatory agencies want to protect your neighbors from interference from your computer. With a vehicle, we must protect ourselves from our computer, too. We cannot shut off vehicle devices if we experience interference, like we can in our homes.

Figure 3 is an example of a typical component radiated emissions test setup. The items to note from this particular setup are whether it is an automotive, consumer, computer, or wireless device, the test setup is essentially a standard format for all radiated emissions testing. As shown in figure 10.3 - the device under test located in one portion of the test chamber, the receiving antenna at the other.

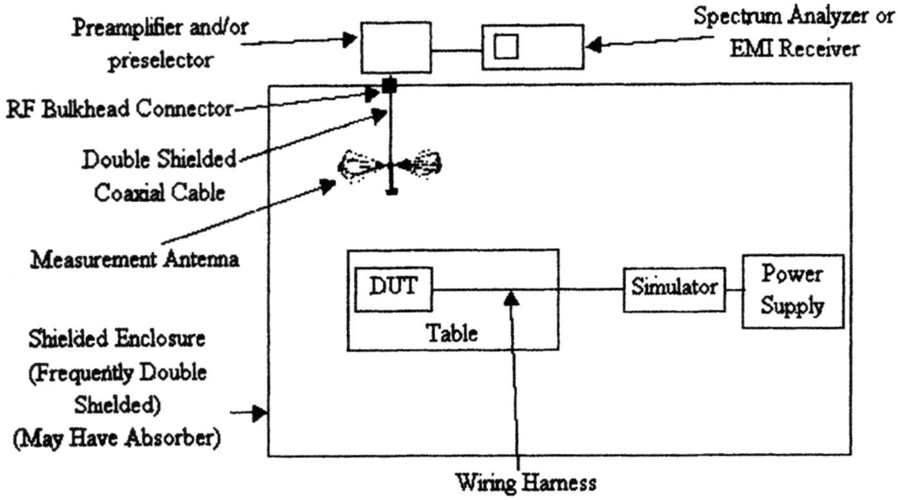

Figure 10.3. **A Typical Radiated Emissions Component Test Setup**

This receiving antenna will be located at some standard distance from the floor or from the walls of the chamber itself. The antenna would then be connected to some type of spectrum analyzer or an EMI receiver to measure the emission coming from the device under test. Many times there are materials covering the wall to absorb reflections. Then the chamber is referred to as anechoic chamber. Sometimes testing is performed in a "reverberant" room." (As a side note, the reverberant room typically yields higher emissions than the anechoic room, because the energy would be reflected from one surface to another and eventually captured by the antenna in the reverberant room. This means measurements may be possible nearer the noise floor.)

The following is an example of the type of information represented in FCC Part 15 is the emission levels are allow for frequencies and their services that are allocated to the frequencies in United States. For example, Table 10.2 shows that at frequencies near 50 MHz, the permitted use, maximum field strength, measurement method, and applicable section of Part 15.

Table 10.2 shows that for approximately the 46 to 50 MHz band, there are many uses that have been allocated by the FCC, and these include items such as cordless telephones and sometimes may include "baby monitors." Also note that there are emission levels that are to be maximums at the certain distance from the antenna, and FCC Part 15 seeks to protect users of the frequencies that limit emissions at the frequencies the certain levels. This essentially is what Part 15 tries to do, and again this is not of particular importance to the automotive industry.

10.5 "MICROVOLTS PER METER" AND WATTS

Watts are the units used to describe the amount of power generated by a transmitter. Microvolts per meter (μV/m) are the units used to describe the strength of an electric field created by the operation of a transmitter.

A particular transmitter that generates a constant level of power (Watts) can produce electric fields of different strengths (μV/m) depending on, among other things, the type of transmission line and antenna connected to it. Because it is the electric field that causes interference to authorized radio communications, and since a particular electric field strength does not directly correspond to a given level of transmitter power, most of the Part 15 emissions limits are specified in field strength.

Although the precise relationship between power and field strength can depend upon a number of additional factors, a commonly-used equation to approximate their relationship is:

$$(PG) / (4\pi D^{2)} = (E^2) / (120\pi)$$

Where

P = the transmitter power in Watts

G = the numeric gain of the transmitting antenna relative to an isotropic source

Table 10.2 **FCC Frequency Table for Frequencies Near 50 MHz**

	Any	100 μV/m @ 3 m	Q	15.209
44.49-46.6 MHz	Any	100 μV/m @ 3 m	Q	15.209
46.6-46.98 MHz	Cordless Telephones	10,000 μV/m @ 3 m	A	15.233
	Any	100 μV/m @ 3 m	Q	15.209
46.98-48.75 MHz	Any	100 μV/m @ 3 m	Q	15.209
48.75-49.51 MHz	Cordless Telephones	10,000 μV/m @ 3 m	A	15.233
	Any	100 μV/m @ 3 m	Q	15.209
49.51-49.66 MHz	Any	100 μV/m @ 3 m	Q	15.209
49.66-49.82 MHz	Cordless Telephones	10,000 μV/m @ 3 m	A	15.233
	Any	100 μV/m @ 3 m	Q	15.209
49.82-49.9 MHz	Any	10,000 μV/m @ 3 m	A	15.235
	Cordless Telephones	10,000 μV/m @ 3 m	A	15.233
49.9-50 MHz	Cordless Telephones	10,000 μV/m @ 3 m	A	15.233
	Any	100 μV/m @ 3 m	Q	15.209
50-54 MHz	Any	100 μV/m @ 3 m	Q	15.209
54-70 MHz	Non-Residential Perimeter Protection Systems	100 μV/m @ 3 m	Q	15.209
70-72 MHz	Intermittent Control Signals	1,250 μV/m @ 3 m	A or Q	15.231
	Periodic Transmissions	500 μV/m @ 3 m	A or Q	15.231
	Non-Residential Perimeter Protection Systems	100 μV/m @ 3 m	Q	15.209
72-73 MHz	Auditory Assistance Devices	80,000 μV/m @ 3 m	A	15.237
	Intermittent Control Signals	1,250 μV/m @ 3 m	A or Q	15.231
	Periodic Transmissions	500 μV/m @ 3 m	A or Q	15.231
	Any	100 μV/m @ 3 m	Q	15.209

14

D = the distance of the measuring point from the electrical center of the antenna, in meters

E = field strength in volts per meter

$4BD^2$ is the surface area of the sphere centered at the radiating source whose surface is D meters from the radiating source. 120π is the characteristic impedance of free space in ohms.

Using this equation, and assuming a unity gain antenna (G=!) and a measurement distance of 3 meters (D=3), a formula for determining power given field strength can be developed:

$$P = 0.3\ E^2$$

Where:
P is the transmitter power (EIRP) in watts, and
E is the field strength in volts per meter.

Chapter 11

Vehicle System Electrical Transients

11.1 BACKGROUND

Automotive systems operate in demanding environmental conditions, have exacting requirements for performance, and are expected to provide years of trouble free operation. They are also expected to be inexpensive and easy to assemble into a vehicle. These systems employ electrical, electro-mechanical, and electronic component technologies that are not used together in other applications. The combination of high voltage and current requirements for some devices can cause problems with the low-level signal or power characteristics of other devices, if precautions have not been taken to isolate these systems.

The area of vehicle system electrical transients represents one of the more challenging areas for the automotive systems designer. This is because many times there is not enough information available on time manner to prevent unanticipated conditions. This chapter discusses some of the basics that need to be considered to minimize issues relating to transient conditions.

11.2 OVERVIEW OF THE VEHICLE TRANSIENT ENVIRONMENT

Studies show that transients that are generated on a vehicle can be as large as 5 to 10 times the supply voltage (which is approximately 13 volts DC). This problem is expected to be exacerbated when vehicles move to higher voltage systems (such as the 42 volt systems presently being developed). This, combined with the increasing number of electromechanical actuators that are anticipated, stresses that there are issues to consider.

There are a number of considerations in the characterization of vehicle transients. First is the duration of the transient, second is the waveform characteristics, and third is the energy content of the transient. Different waveforms may have similar voltage characteristics, yet different energy content. The final consideration is how frequently the transients occur during vehicle operation. The most severe transients are caused by the switching of inductive loads, which occurs frequently. These transients can induce voltage or current in wires within an adjacent harness. Studies have shown that these induced transients can be as high as a few hundred volts.

11.3 COMPONENT SELECTION

The basis for any electrical design is the components. Selection of components for EMC characteristics is equally as important as selection of components for their intended function. Except for wideband video and circuits employing oscillators, analog circuits are generally much quieter than digital circuits. Because digital circuits tend to be noisier, this section emphasizes the selection of digital components for suppression of EMI. The most important issue in selecting digital components for low-noise characteristics is rate of change of energy. The noise voltage induced into a victim circuit from a noise source circuit is:

$$V = -M\, dI/dt$$

Equation **11.1**

where M is the mutual inductance between the two circuits and the coupling is magnetic. See Figure 11.1.

Or:

$$V = C\, dV/dt$$

Equation **11.2**

where C is the capacitance between the two circuits. Coupling is electric in nature. See Figure 11.2. Mutual inductance, M, depends on current loop areas of source and victim, relative loop orientation, separation distance, and the heights of the circuits above ground. Source and victim current loops are analogous to the primary and secondary windings of a transformer. Capacitance, C, depends on the distance between conductors, associated effective areas, and Z, the impedance to ground of the victim circuit. The source and victim conductors act as a parallel plate capacitor.

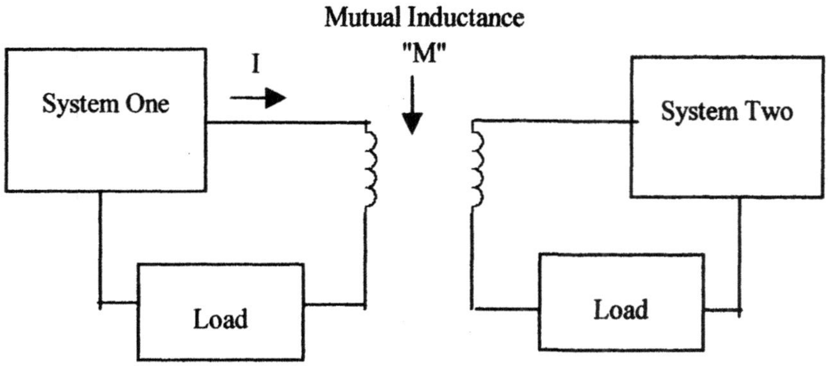

Figure 11.1. **Noise Coupling via a Primarily Magnetic Field**

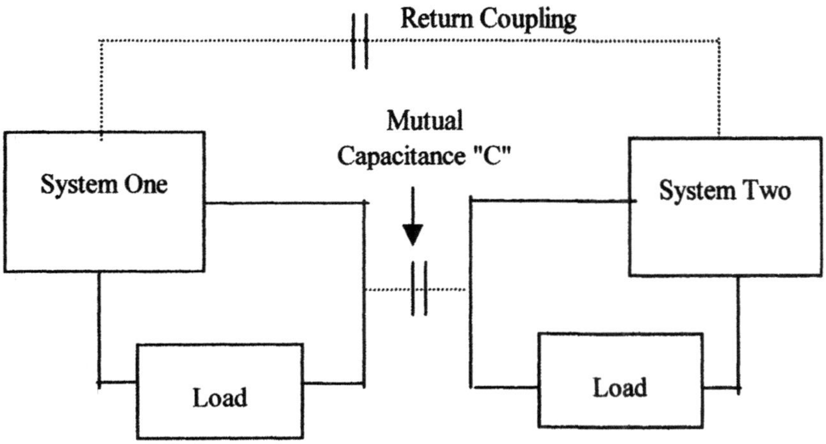

Figure 11.2. **Noise Coupling via a Primarily Electric Field**

11.4 LOGIC FAMILIES AND D*V/DT*

Table 11.1 shows various digital family rise times and voltage rates of change *(dV/dt)*. The faster the rise time and the higher the voltage swing, the larger the d*V/dt*. Using the slowest rise time to achieve the desired function can reduce the amount of noise coupling. Another reason for using slower rise time is to limit the higher frequency harmonics of the digital signal to minimize RE.

Because the circuit traces on printed circuit boards (PCBs) can act as antennas and radiate noise at higher frequencies, limiting the unnecessary harmonics in a digital signal prevents radiation of these higher frequency harmonics.

Table 11.1. **Rise Time and Voltage Rate of Change for Various Logic Families.**

Logic Family	Rise Time (ns)	Voltage Swing (V)	d*V/dt* (V/ns)
CMOS 5 V	100	5	0.5
CMOS 12 V	25	12	0.48
CMOS 15 V	50	15	0.30
HCMOS	10	5	0.5
TTL	10	3	0.3
ECL 10k	2	0.8	0.4
ECL 100k	0.75	0.8	1.1

11.5 LOGIC FAMILIES AND *DI/DT*

As a result of stacking the output stage of the logic circuit in the chip (Figure 11.3), when the logic is switched, the transistors typically turn off slower than they turn on and draw large amounts of transient current from *Vcc* during the turn-off transition. This induces transients on the Vcc trace and return. Notice that the output stage of the TTL circuit in Figure 11.3 contains a current-limiting resistor. The CMOS circuit has no current-limiting resistor and, consequently, draws larger currents *(dI/dt* sometimes as high as 5,000 A/s) than TTL. One way to limit these surges is through the use of decoupling capacitors. The decoupling capacitor, which will supply the necessary instantaneous currents while the chip is switching, is a capacitor connected between *Vcc* and return (Figure. 11.4). It is important to remember to make the capacitor leads as short as possible to reduce parasitic inductance, and to mount the capacitor close to the decoupled chip to reduce loop area.

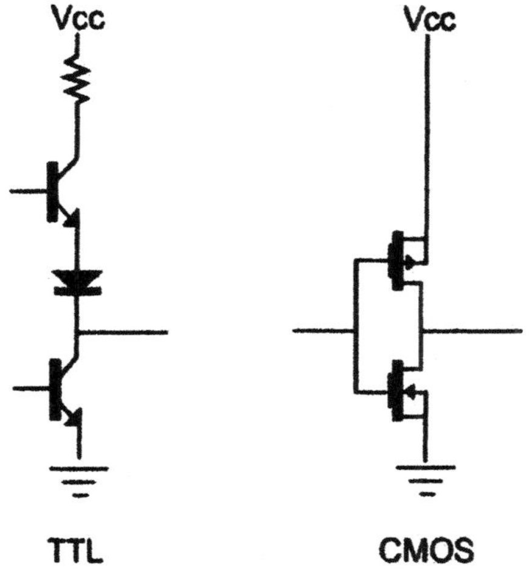

Figure 11.3. **Logic Output Drivers**

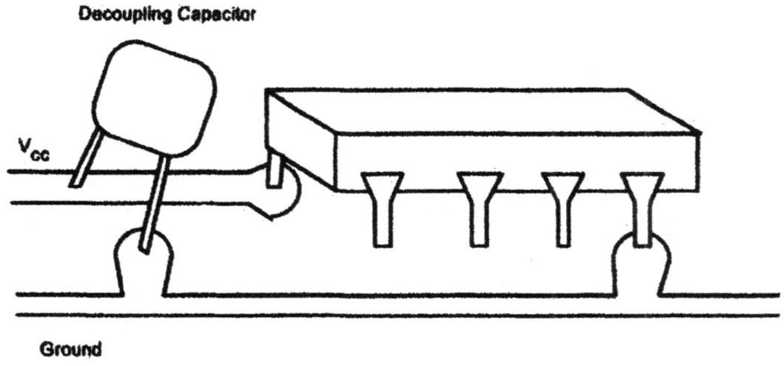

Figure 11.4. **IC Chip and Decoupling Capacitor**

11.6 LOAD INDUCED SWITCHING TRANSIENTS

Figure 11.5 portrays the elements of a power distribution system. It consists of a power source, wiring, and a load. The power source is represented by an ideal voltage source in series with a resistor and/or inductor. The wiring contributes both resistance and inductance. The load, at turn-on or turn-off, attempts to generate a rapid change of current through the power source and wiring impedance. Load capacitance is a short circuit to ground at the instant it is connected to the bus. The distribution wiring inductance impedes this current demand by generating a potential of such polarity as to oppose the flow of new current. The resultant waveforms are depicted in Figures 11.6 and 11.7. This simple model ignores any capacitive effects, other than the load. Source parallel capacitance (especially in a dc supply) contributes to low source impedance, which may be easily modeled in the transient case by using a smaller series source impedance. Line-to-line or line-to-return wiring capacitance is easily accounted for by modeling the distribution wiring as an inductance in series with a resistor. That is to say, a lumped - element model of a transmission line, otherwise known as a line impedance stabilization network (LISN). Figure 11.8 shows a model for both calculated and measuring switching transients. In Figure 11.8, the LISN simulates the distribution wiring impedance. The heavy lines show the flow of high current to the spike-generating load.

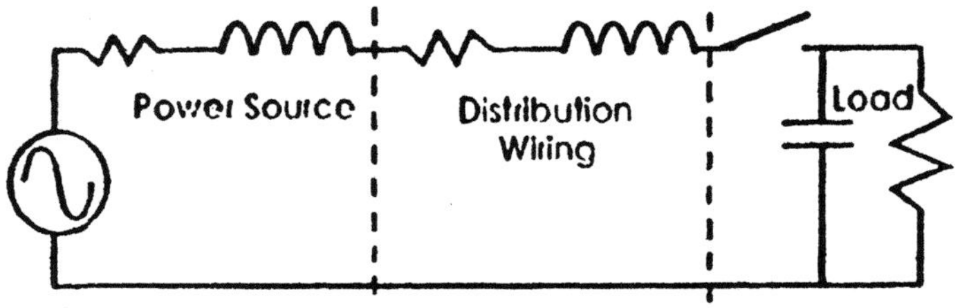

Figure 11.5. **Model of Electrical Power Distribution System**

It is important to have a complete physical description of the phenomenon of switching transients. This includes waveform amplitude vs time envelopes, as well as source impedance. It is also important to have accurate repeatable measurements of load-induced switching transients.

The energy that is stored in an inductive device is determined from

$E = \frac{1}{2} LI^2$

Where E = energy in joules,

L = inductance, and

I = steady state current

Equation **11.3**

This is the energy that is responsible for the generation of the "turn off" transient.

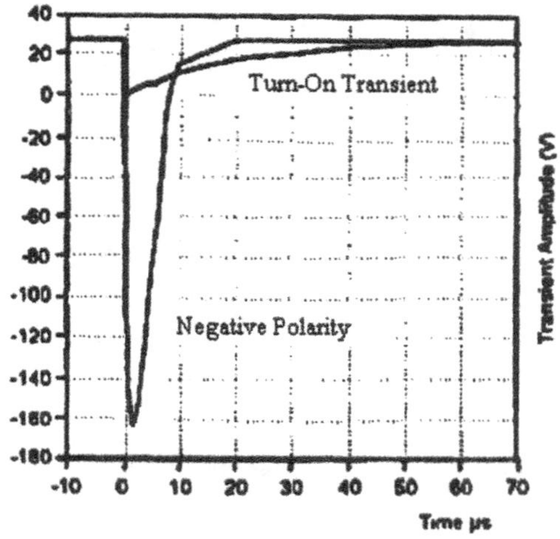

Figure 11.6. **Turn-On Switching Transient**

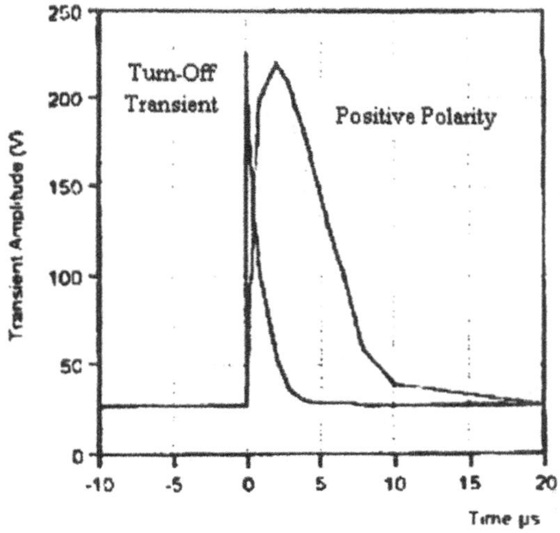

Figure 11.7. **Turn-Off Switching Transient**

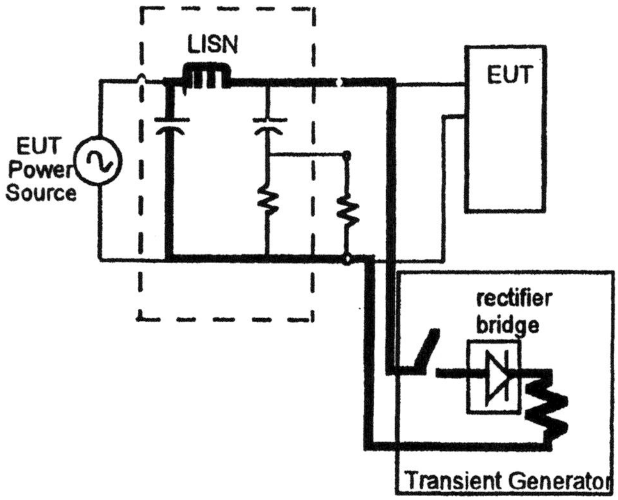

Figure 11.8. **Spike Generator Model**

11.7 SPECIFYING CONTROL OF THE SWITCHING TRANSIENT PHENOMENON

From the previous discussion, we see that switching transient magnitude (amplitude, time duration) can be estimated from a knowledge of bus characteristics, including nominal bus potential and bus impedance. The following facts are immediately discernable.

These transients drop nearly instantaneously from nominal bus potential towards zero potential, then gradually return to nominal following a (typically) underdamped exponential decay. The generic shape is that of the negative going long duration transient. The excursion never dips below zero. The time to return to nominal depends on the reactive elements of the switched load (mainly line-to-line capacitance) and the reactive and resistive impedance of the power bus.

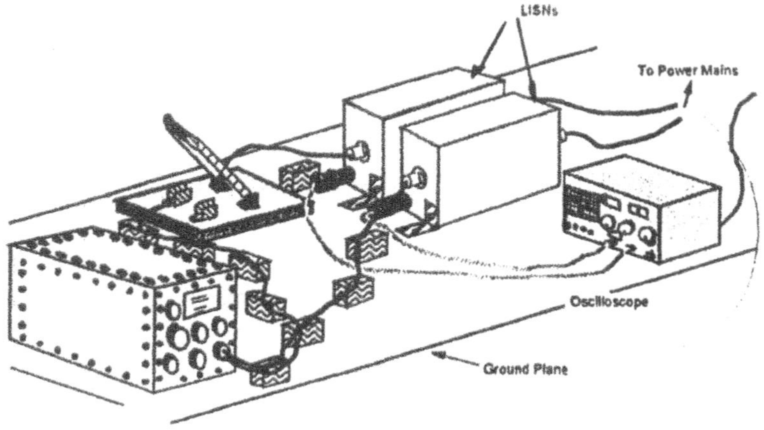

Figure 11.9. **Transient Measurement Setup**

11.8 METHODS TO MINIMIZE THE IMPACT OF TRANSIENTS

When the systems have been incorporated into a vehicle, and the appropriate actions have been taken to minimize transient generation, yet transient signals still exist, the last resort may be to suppress the transients at the receiver connection. These suppression methods would typically consist of a component that would absorb the transient energy present.

The transient absorbing components may be of two types; passive devices or active devices. Each of these has its own characteristics that would make it a candidate for specific applications.

The passive devices would consist of resistors or capacitors. The resistor would minimize the transient by dissipating the energy in the form of heat and preventing that energy from affecting the component that is being protected. The capacitor may be used in a mode that would allow it to absorb the voltage spikes and minimize their amplitude by presenting lower impedance to the transient than the impedance of the protected component. The capacitor would not dissipate the energy, it would reduce the amplitude of the voltage, store it for some amount of time, and then discharge into the load at a much slower rate than the original transient.

Active devices consist of essentially diode devices, and in some ways these would operate similar to the capacitors. They would present low impedance to minimize the voltage on the load. They may dissipate some energy, so they need to be rated for their application.

11.9. TRANSIENT SUPPRESSION CIRCUIT TOPOLOGIES

The last item to be addressed is the specific circuit configuration that would be used. There are two types, the "centralized" and the "distributed" methods.

The centralized suppressor is placed near the source of the transient, and the distributed suppressor is placed near the receiver of the energy (each device that is being protected). Placing the suppressor at the source eliminates the need for many suppressors located locally throughout the vehicle. This may mean that the suppressor will need to be large, since it is intended to suppress all the source energy. The advantage of distributed suppression is that each protected device will have local suppression, although suppression may be complicated to put on each device.

11.10 CONCLUSIONS

Transients on vehicles result from many different types of loads and circuits that may have fast energy rise and/or fall times. These transients can be difficult to anticipate, since they also depend upon the load and source impedances of the system.

It may be possible to use transient suppression devices, though there are trade offs with the selection and type of these devices.

Chapter 12

Electrostatic Discharge

12.1 OVERVIEW OF ESD

A discussion of electrostatic discharge (ESD) is typically introduced later in the study of most texts on EMC, and this one is no exception. This is because to understand ESD, we need to know the characteristics of both conducted and radiated phenomena discussed in earlier portions of the material. ESD can have emissions that affect devices or components that do not have sufficient immunity.

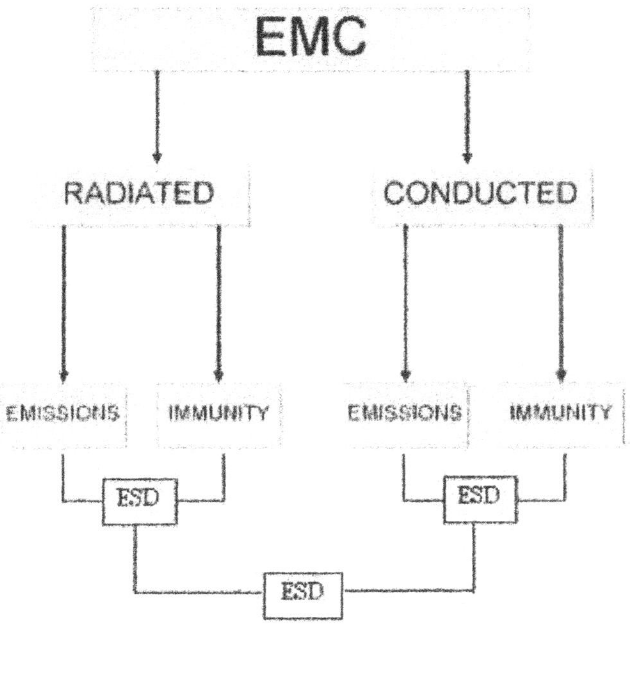

Figure 12.1 **ESD Can Affect Radiaed or Conducted Phenomena**

The basic mechanism of ESD is that the capacitance between two surfaces can store a charge, resulting in a voltage difference between the surfaces. A discharge may occur when the voltage between the surfaces is greater than the dielectric capability of the dielectric material between the surfaces. Many times the discharge takes place in air, although it can occur across any dielectric material.

Why is ESD of concern today? We need to review the evolution of the electronics industry. If we look at the early days of electronics, and focus on the time where radio and television communication were becoming common, most of the active electronic elements were vacuum tubes. One of the characteristics of vacuum tubes is that they needed hundreds of volts to operate, and typically a thousand volts would be required to cause damage.

As technology evolved, semiconductor devices replaced vacuum tubes in many applications. One benefit of semiconductor devices is that they operate on lower voltages (in the case of transistors, tens of volts). As an undesired consequence, damage can occur to semiconductor devices with only hundreds of volts.

The schematic symbols for a vacuum tube, transistor, and diode are shown in figure 12.2.

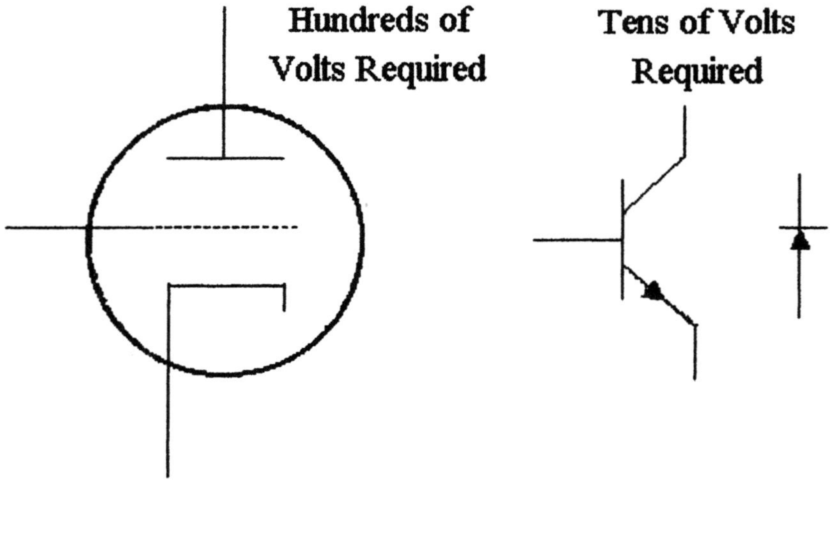

Figure 12.2. **Vacuum Tube and Solid State Devices**

As technology has continued to increase the level of device integration, less and less voltage is required to operate the devices, with a reduction in the amount of voltage required to cause damage.

12.2 THE ROLE OF INSULATING MATERIAL IN ESD

The generation of an ESD event is caused by an accumulation of opposite charges on two surfaces. When these surfaces have a distance between them, a potential voltage difference is developed. Key to this is that the surfaces are insulators or are isolated from each other by an insulating material. Without the insulating material, the charges would distribute themselves across both materials and equalize the charges so no potential difference remained. Materials that are normally considered insulators in most circuits (with perhaps tens of thousands of ohms) are actually effective conductors to eliminate potential ESD conditions.

An example of this is the conductive wrist strap and return connection that is used to prevent the build up of charge on a person in an electronics assembly process. The resistance of the wrist strap and connecting cable is relatively high, yet provides enough conductivity to eliminate a charge build-up.

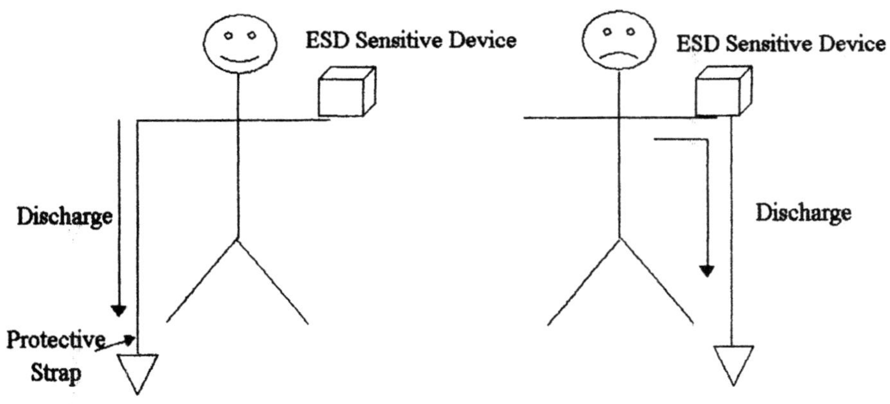

Figure 12.3. **ESD Protective Strap Functionality**

The equation that relates charge, voltage, and capacitance is:

$$Q = CV$$

Where C is the capacitance between two surfaces,
V is the voltage that is developed between the two surfaces,
and Q is the charge on either surface.

From this relationship, it can be seen that:

$$V = Q / C$$

If the charge stays constant (meaning no charge is being drained away) and the capacitance decreases, the voltage will build up between the two surfaces. As the voltage builds, there will be a point where the breakdown voltage is achieved, and the discharge occurs. This voltage breakdown occurs at approximate 75 kV / inch in air.

This is shown in figure 12.4.

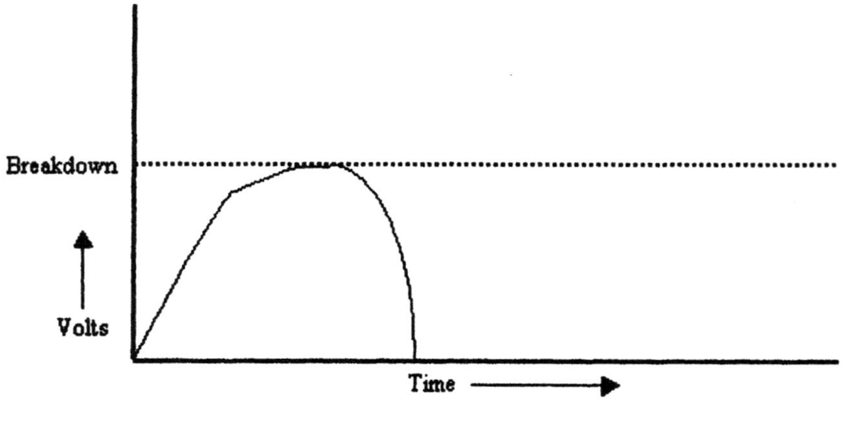

Figure 12.4. **ESD Voltage Breakdown**

This is important to keep in mind when determining the ESD requirements for electronic devices handled by people. We can model this situation in the following manner. Recall from basic physics that two surfaces have a capacitance between them. If one of the surfaces was at a distance of infinity, we would say that we have the "free space capacitance" of the remaining surface. This is the basis for modeling the equivalent capacitance of people. We are concerned about the charge that can be built up by the body because it may affect electronic devices. Empirical research has suggested that the equivalent free space capacitance for the human body is from 50 to 150 pf, and varies as a function of the surface area of the person. Just as with a capacitor, as the surface area increases, the capacitance increases.

12.3 HUMAN BODY MODEL FOR ESD

In addition to the equivalent free space capacitance, the complete human body model (HBM) used in ESD work includes resistance and inductance. The model can be represented as a series RLC network, as shown in figure 12.4.

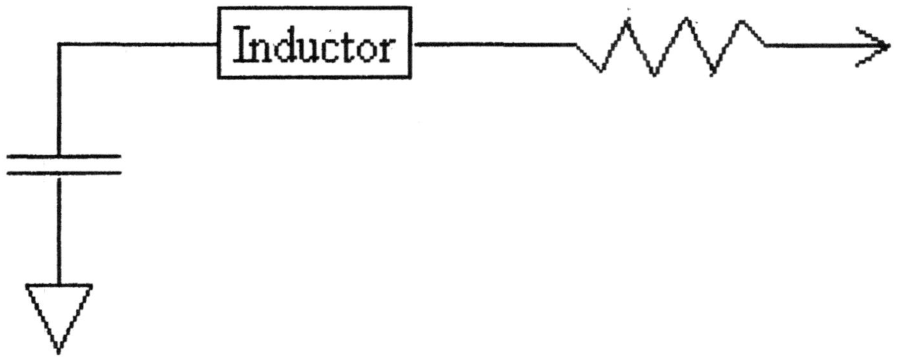

Figure 12.5 **Human Body Model for ESD**

Typical values for the inductance are a few tenths of a μH. The values of the resistance are dependent upon the skin characteristics and range from tens of Kilo ohms to hundreds of Kilo ohms.

12.5 EFFECTS OF ESD

The discharge can induce damage to the device. The challenge here is the two types of damage that can occur. The first one is immediate, and the second one is termed "latent". Each has its own characteristics, advantages, and disadvantages.

Table 12.1 **Characteristics of Types of Failures:**

	Immediate	**Latent**
Advantage	Easily Detectable	Device May Continue to Operate
Disadvantage	Causes Immediate Failure	Device May Fail Later

12.6 ESD TEST METHODS

Automotive EMC programs have ESD requirements. These requirements include both component and system-level requirements. Testing is conducted by simulating an ESD event occurring at or near the device. This is accomplished by use of an ESD "gun," as shown in Figure 12.5 for a typical component level ESD test set-up.

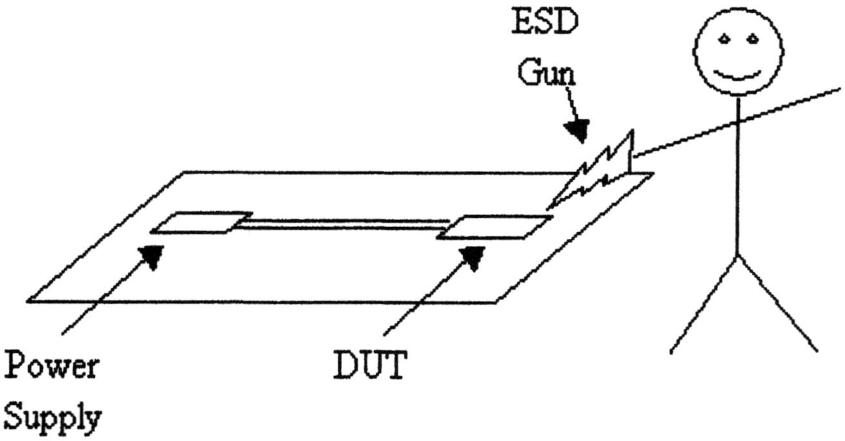

Figure 12.6. **ESD Test Setup**

At a system level, ESD testing would be conducted by applying the high voltage to various wires and connections on the system itself. The intention is to provide enough charge to activate any device failure. The advantage of system level testing is that, due to the connection of several devices and wiring harnesses, the charges are distributed and minimized so the devices do not fail.

ESD test conditions need to be specified in detail, because the charge and voltages that are applied depend upon environmental conditions such as temperature and humidity.

In summary, ESD is a unique aspect of EMC, yet devices and auto systems can be tested to simulate the effect of ESD.

APPENDIX A

ACRONYMS AND ABBREVIATIONS

A	ampere
A/m	ampere per meter
ac	alternating current
AM	amplitude modulation
AWG	American wire gauge
BB	broadband
BCI	bulk current injection
C	capacitor
CE	conducted emissions
cm	centimeter
CM	common mode
CI	conducted immunity
CUT	cable under test
CW	continuous wave
dB	decibel
dBm	decibel above 1 milliwatt
dBV	decibel above 1 volt
dBW	decibel above 1 watt
dc	direct current
DSO	digital storage oscilloscope
DM	differential mode
ELF	extremely low frequency
EHF	extremely high frequency
EM	electromagnetic
EMC	electromagnetic compatibility
EME	electromagnetic environment
EMI	electromagnetic interference
EMP	electromagnetic pulse
ESD	electrostatic discharge
EUT	equipment under test
FFT	fast Fourier transform
FM	frequency modulation
GHz	gigaHertz
H	Henry
HF	high frequency

Hz	Hertz
I/O	input/output
kHz	kiloHertz
km	kilometer
IF	intermediate frequency
L	inductor
LC	inductive/capacitive
LH	left hand
LISN	line impedance stabilization network
LF	low frequency
LO	local oscillator
m	meter
MF	medium frequency
MHz	megaHertz
mm	millimeter
N/A	not applicable
NASA	National Aeronautics and Space Administration
NB	narrowband
pF	picoFarad
PM	phase modulation
PC	personal computer or printed circuit
PCB	printed circuit board
PRF	pulse repetition frequency
PAM	pulse amplitude modulation
PCM	pulse code modulation
PWM	pulse width modulation
RAU	remote acquisition unit
RBW	resolution bandwidth
RC	resistive/capacitive
RCVR	receiver
RE	radiated emissions
RF	radio frequency
RFI	radio frequency interference
RH	right hand
rms	root-mean-square
RI	radiated immunity
s	second
S/N	signal-to-noise ratio
SHF	super high frequency
SMPS	switched mode power supply
T	tesla
TT	turn-on/off transient
UHF	ultra high frequency
V	volt

V/m	volt per meter
VBW	video bandwidth
VF	voice frequency
VHF	very high frequency
VLF	very low frequency
VTVM	vacuum tube volt meter
W	watt
XFMR	transformer
ε (epsilon)	permitivity
λ (lambda)	wavelength
μ (mu)	permeability or prefix micro
Ω (omega)	ohm
μ F	microFarad

APPENDIX B

USEFUL FORMULAS

Receive Antenna Factor

$AF = E/V$
Where

AF = Antenna Factor in 1/meters
E = Field Strength, V/m or μV/m

Converting to dB notation:

$AF\ (dB/m) = 20 \log (E/V)$
Or
$AF\ (dB/m) = E\ (dBV/m) - V\ (dBV)$

The antenna factor may be calculated from:
$AF = 9.73/(\lambda \sqrt{g})\ (m^{-1})$
Where
λ = Wavelength in meters
g = numeric antenna gain

Antenna factors for loop antennas used to measure magnetic fields:
$AF_{H\ dB\ (S/m)} = H_{dBA/m} - V_{dBV}$

In terms of flux density (B Field)
$AF_B = AF_H + 20 \log \mu$
$AF_B = AF_H - 118$, T/V

For loop antennas calibrated in terms of equivalent far field electric field,

$AF_{E\ dB(m-1)} = AF_{H\ dB\ (S/m)} + 20 \log \eta$
$AF_{E\ dB(m-1)} = AF_{H\ dB\ (S/m)} + 20 \log (120\pi)$, or
$AF_{E\ dB(m-1)} = AF_{H\ dB\ (S/m)} + 51.5$ dB
Where
$\eta = 120\pi\ \Omega$ = impedance of free space

Conversion of levels from mW to μV (in a 50 Ω system:)
$P = V^2/R$

Where
P = Power in Watts
V = Voltage in Volts
R = Resistance in Ω

For power in milliwatts and voltage in microvolts,

$V_{dB\mu V} = P_{dBm} + 107$

Power Received by an antenna, transmitted by another antenna
$Pr = P_t G_t G_r \lambda^2 / (4\pi r)^2$
Where
Pr = Power received, Watts
P_t = Power transmitted, Watts
G_t = Numeric gain of transmit antenna
G_r = Numeric gain of receive antenna
r = Distance between antennas in meters
λ = Wavelength of transmitted signal in meters

Electric field strength in the far field from a transmitter
$E_{V/m} = (\sqrt{(30 P_t G_t)})/r$
Where the terms are defined in 4.5

To solve for power required to generate a specific field, the equation can be rewritten as
$P_t = E^2 r^2 / 30 G_t$

For example, if we need to generate 30 V/m at 500 MHz from an antenna three meters away, and the antenna has a numeric gain of two, the required power at the antenna (amplifier output power minus path loss) would be:

$P_t = (30)^2 (3)^2 / 30(2) = 135$ Watts

For low gain transmitting antennas, far field conditions exist at
$r \geq \lambda / 2\pi$
For high gain transmitting antennas, far field conditions exist at
$r \geq 2D^2 / \lambda$
where
D = Maximum dimension of antenna in meters

Relationship of antenna factor and gain for a 50 Ω antenna
$G_{dB} = 20 \log (f_{MHz}) - AF_{dB/m} - 29.79$

Power required to generate a desired field strength at a given distance when antenna factors are known
$P_{dB(W)} = 20 \log(E_{desired\ V/m}) + 20 \log(d_m) - 20 \log(f_{MHz}) + AF_{dB/m} + 15$

Transmit antenna factor
$TAF_{dB/m} = G_{dB} - 2.22 - 20 \log (d_m)$
where
$TAF_{dB/m}$ = Transmit antenna factor in dB/m
G_{dB} = Antenna gain in dB
D_m = distance in meters

List of equivalent magnetic units
$\mu_0 = 4\pi(10)^7$ H/m

1 T	1 W/m^2	$7.96(10)^5$ A/m	$(10)^4$ G	$(10)^4$ gamma

1 nT	796 A/m	$(10)^3$ pT	$(10)^{-5}$ G

1 μA/m	$1.256(10)^{-3}$ nT	1.256 pT	$1.256(10)^{-8}$ G	$1.256(10)^{-3}$ gamma

1 pT	0.796 μA/m	$(10)^{-3}$ nT	$(10)^{-8}$ G	$(10)^{-3}$gamma

1 gamma	796 μA/m	1 nT	$(10)^{-3}$ pT	$(10)^{-5}$ G

1 oersted = 1 Gauss
0 dBμA/m = +2 dB(pT)

Gauss, oersted, and gamma are obsolete terms

1- EM Characteristics:

a) $f \times \lambda = c$
b) $f \times h = E$

f = frequency (cycles/sec or Hertz)
λ = wavelength (m)
c= speed of light = 3 (10^8) m/sec
h= Planck's Constant = 6.626 W 10^{-27} erg-sec. = 4.13 W 10^{-16} eV-sec
E= Photon energy in ergs or eV

2- Power Density:

a) $PD = E \times H$
b) If $E \perp H$,

E/H = 377 Ω, and
$PD = E^2/3770 = 37.7\ H^2$

PD = Power Density (in Watts/cm^2)
E = Electric Field (volts/meter)
H = Magnetic Field (Amps/meter)
X = Vector Cross Product
ee space impedance

3 Antenna Equations

a) $W_{nf} = (4P) / (A)$
b) $W_{ff} = (AP) / (\lambda^2 r^2)$
c) $R_{ff} = (A) / (2\lambda)$
d) $R = (PG)^{1/2} / (\pi EL)$
e) $G = (4\pi A) / (\lambda^{2)}$

Wnf = Max Power Density in the near field of the antenna
P = Power Output (Watts)
A = Area of antenna (m^2)
W_{ff} = Maximum Power Density in the far field of the antenna
Rff = distance to far field
λ = Wavelength of radiation (meters)
r = Distance to specified power density ((mW) / (cm^2))
EL = Specified power density ((mW) / (cm^2))
G = Antenna gain (numerical, not dB)

References:

American National Standards Institute (ANSI), "American National Standard for Methods of Measurement of Radio-Noise Emissions From Low Voltage Electrical and Electronic Equipment in the Range of 9 kHz to 40 GHz", ANSI C63.4, 1992

"Antennas and Radio Propagation", Department of the Army Technical Manual, TM 11-666, United States Department of the Army, February, 1953

B. Archambeault, "PCB Design for Real-World EMI Control", Kluwer Academic Press, 2002

"The ARRL Antenna Book", 19th Edition, The American Radio Relay League, Inc., 2000

G. Brown, "And Part of Which I was – Recollections of a Research Engineer", Hanlo Press, 1982

Compliance Engineering Magazine, January / February 2001

Department of the Army Technical Manual TM 11-483, Department of the Air Force Technical Order, TO 31-3-9, "Radio Interference Suppression", United States Departments of the Army and the Air Force, October, 1956

"Designing for EMC", Presented at the IEE colloquium "EMC for the Small Business", May 1997 at the IEE, Savoy Place, London, http://www.elmac.co.uk/desfremc.htm

Electronics safety pages, Electromagnetic Compatibility (EMC), http://www.epanorama.net/links/safety.html#emc

"EMC: Antennas and Antenna Factors, How to Use Them", Society of Automotive Engineers AIR 1509, SAE, 1980

ETS Lindgren

Federal Communications Commission (FCC), "Bulletin Number 63, Understanding The FCC Part 15 Regulations for Low Power, Non-Licensed Transmitters", Federal Communications Commission

Federal Communicaitons Commissions (FCC), "Part 15", Federal Communications Commission

Field service memo
Electromagnetic radiation and how it affects your instruments

"F-M Transmitters and Receivers", Department of the Army Technical Manual, TM 11 – 668, United States Department of the Army, September, 1952

J. Goebloed, "Traceability and Accuracy Aspects in EMC Compliance Testing", EMC '94 Roma, International Symposium on EMC, September 13-16 1994, Rome

J. Grady, "System Requirements Analysis", McGraw-Hill, 1993.

E. Hare, Editor, "The ARRL RFI Book", First Edition, The American Radio Relay League, 1998

W. Hayt, "Engineering Electromagnetics", McGraw-Hill, 1974.

D. Heirman, "EMC in a High-Frequency World", Compliance Engineering Magazine, Jan-Feb 2001

IEC, "Publication 801 Part 4, Electromagnetic Compatibility for Industrial Process Measurement and Control Equipment, Part 4: Electrical Fast Transient/Burst Requirements", IEC, 1988

"Introduction to wave propagation, transmission lines, and antennas", Book 10, Integrated Publishing, 2003, http://www.tpub.com/neets/book10/index.htm

Integrated Publishing, DOE Electrical Science, http://www.tpub.com/doeelecscience/

G. Johnson, Jr.., "Electromagnetic Interference Control Guidance for the Equipment Designer", United States Army Signal Research and Development Laboratory, Fort Monmouth, New Jersey, February 1, 1961.

J. Kraus, R. Marhefka, "Antennas for All Applications", Mc Graw Hill, 2002

R.D. Leach, M.B. Alexander, Editor, "NASA Reference Publication 1374, Electronic Systems Failures and Anomalies Attributed to Electromagnetic Interference", July, 1995

Marconicalling.com, http://www.marconi.com/html/homepage/home.htm

J. Muccioli, "Electromagnetic Standards and Interference", Automotive Electronics Handbook, Second Edition, McGraw Hill Companies, Inc., 1999

J. Muccioli, "Electromagnetic Compatibility", Automotive Electronics Handbook, Second Edition, McGraw Hill Companies, Inc., 1999

NASA MIL-B-5087B, "Bonding, Electrical, and Lightning Protection, for Aerospace Systems", April 1997

NASA / CR – 1999-209574, "Specification, Measurement, and Control of Electrical Switching Transients", Marshall Space Flight Center, September, 1999

NASA Reference Publication 1368, "Marshall Space Flight Center Electromagnetic Compatibility Design and Interference Control (MEDIC) Handbook", CDDF Final Report, Project No. 93-15, June 1995

"National Electrical Code", National Fire Protection Association, 2002

J. Osburn, "How to Figure EMC Antenna Factors", Evaluation Engineering Magazine, February 1995

H. Ott, "Noise Reduction Techniques In Electronic Systems", John Wiley and Sons, Inc., 1988

C. Paul, "Introduction to Electromagnetic Compatibility", John Wiley and Sons, Inc., 1992.

"PC-113 Test System", Com-Power Corporation, 1999

"Power Losses and Gains – Decibels", Information Text 53005A, U.S. Army Signal Center and School, 1968

"Radio Interference Suppression Techniques", Suppression and General Engineering Branch Coles Signal Laboratory, Fort Monmouth, New Jersey, November 1953

United States Department of Defense, MIL-F-15733, "Filters and Capacitors, Radio Frequency Interference, General Specification For"

United States Department of Defense, "MIL-STD 462 Electromagnetic Interference Characteristics, Measurement of", July 31, 1967, Notice 2, May 1, 1970

United States Department of Defense, "MIL-STD 461 Electromagnetic Interference Characteristics for Equipment" Basic Release July 31, 1967, Revision A, August 1968, Revision B, Electromagnetic Emission and Susceptibility Requirements for the Control of Electromagnetic Interference", Aprill, 1980, Revision C, August 4, 1986.

University of Missouri – Rolla Electromagnetic Compatiblity Laboratory, http://umr

U.S. Occupational Safety and Health Agency (OSHA), http://www.osha.gov/SLTC/radiofrequencyradiation/electromagnetic_fieldmemo/electromagnetic.html

J. Shen, "MOSTVS: A New Class of Transient Voltage Suppressors to Reduce Voltage Rating and Cost of Automotive Power Electronics" SAE World Congress, 2003

T. Van Doren, "Grounding and Shielding Electronic Systems", T. Van Doren, 1998

T. Van Doren, "Circuit Board Layout to Reduce Electromagnetic Emission and Susceptibility", T. Van Doren, 2000

T. Williams, "EMC and Litigation: some thoughts from the High Court", Elmac Services UK, http://www.elmac.co.uk/zurhcemc.htm

T. Williams, "What to Look for in an EMC Antenna", Compliance Engineering Magazine, 1999 Annual Issue

Index

GPSR Compliance

The European Union's (EU) General Product Safety Regulation (GPSR) is a set of rules that requires consumer products to be safe and our obligations to ensure this.

If you have any concerns about our products, you can contact us on ProductSafety@springernature.com

In case Publisher is established outside the EU, the EU authorized representative is:

Springer Nature Customer Service Center GmbH
Europaplatz 3
69115 Heidelberg, Germany

Batch number: 10755284

Printed by Printforce, the Netherlands